Laboratory Manual for Miller's

Living in the Environment, Environmental Science, and Sustaining the E

C. Lee Rockett
Bowling Green State University

Kenneth J. Van Dellen
Macomb Community College

Wadsworth Publishing Company
Belmont, California
A Division of Wadsworth, Inc.

Science Editor	Jack Carey
Editorial Assistant	Kristin Milotich
Production Editor	Deborah Cogan
Managing Designer	Andrew Ogus
Print Buyer	Randy Hurst
Art Editor	Nancy Spellman
Permissions Editor	Peggy Meehan
Copy Editor	Alan Titche
Technical Illustrators	Barbara Barnett, Marilyn Krieger
Compositor	Vicki Moran
Cover Photos	Top: © Larry Ulrich; bottom: © Pat & Tom Leeson/Photo Researchers, Inc.
Printer	Malloy Lithographing

14 15 16 17 18 19 20 - 08 07 06 05

Printed in the United States of America

ISBN 0-534-17809-X

Contents

Preface

This laboratory manual includes a variety of activities. Some are conventional laboratory exercises, some are more like workbook exercises, and some are more like projects. This is partly because some environmental topics involve processes that take place over considerable time or vast areas, and simply cannot be duplicated or even modeled in the laboratory. Also, use of a variety of approaches adds some interest. Exercises were designed, as much as possible, to use a minimum of sophisticated equipment.

The exercises are organized in basically the same order as the chapters in G. Tyler Miller's *Environmental Science*, but the manual is suitable for use with either that text, Miller's *Living in the Environment*, or *Sustaining the Earth*, or even with others.

In solving environmental problems, we use various areas of science, such as biology, chemistry, geology, physics, and meteorology. These help us to understand the interactions between the different elements of the environment, including ourselves. We also use non-science disciplines, such as economics, politics, and sociology, in order to determine how to minimize our impact on the environment. These exercises draw from several of these science and non-science disciplines.

ACKNOWLEDGMENTS

We would like to thank the following people for their help in preparing or commenting on this manual: Frank Budd, Efficient Sanitation, for data on recycling; Phillip Brown, Greater Detroit Resource Recovery Authority, for data and other information; Sheila Dilbeck, Marion Merrell Dow Inc., for assistance in securing permission to use the smoking survey; Dale Krajniak, Manager, City of Grosse Pointe Park, Michigan, for help with data on recycling; Patrick Dailey, Lewis and Clark College; Lee A. Floro, Bowling Green State University.

READING LIST

Scientific American, September 1990

Booklets on energy from: Thomas Alva Edison Foundation
3000 Book Building
Detroit, MI 48226

James T. Dulley, *60 Most Popular* Cut Your Utility Bills *Columns and Updates*

James T. Dulley, *Alternatize Your Home*
   Starcott Media Services
6906 Royalgreen Drive
Cincinnati, OH 45244

1   Introduction to the Compound Microscope

EQUIPMENT NEEDED

1. Compound microscopes
2. Letter "e" slides
3. Colored thread slides
4. Assorted slides of instructor's choice

The compound microscope is an indispensable tool for viewing very small objects, organisms, and parts of larger organisms too small to see with the naked eye. A microscope is an expensive, precision instrument. Your success and enjoyment in many of the labs in this course will depend on your ability to use it correctly. Even if you have had some experience using a microscope before, it is important that you follow these instructions carefully. Do not begin the following exercise without your instructor's supervision. Because each type of microscope is different, it is important that you familiarize yourself with your model.

The compound microscope is designed to magnify thin, semitransparent objects up to approximately 1,000 times (X). It is called **compound** because it has two kinds of lenses, the **ocular** (in the eyepiece; see Fig. 1-1) and the **objective** (close to the object being magnified). Both contribute to the total magnification, which is calculated by multiplying the power of the ocular by the power of the objective. For example, using a 10X ocular and a 43X objective, the total magnification is 430X. This means that an object seen under the microscope will appear 430 times larger than it actually is.

Get a microscope from the cabinet as directed by your instructor. The correct way to carry a microscope is with one hand firmly gripping the arm, and the other supporting the base from below. If you have a built-in light to provide direct illumination, carefully unwrap the cord from the base and plug it in. Using Figure 1-1, familiarize yourself with the names of the other parts of the instrument; please label the diagram. Your microscope may differ slightly in construction from the "model" illustration; your lab instructor will point out the differences.

PARTS OF A COMPOUND MICROSCOPE

Base: Bottom part of the microscope, on which everything else rests.

Arm: Curved metal piece that connects the stage, base, and nosepiece.

Stage: Flat area on which prepared specimen rests. NOTE: The stage should be kept level at all times. Your microscope may have a mechanical stage that moves either vertically or horizontally by means of adjustment knobs.

Stage Clips: Metal clips on stage to keep slide in place. If you have a mechanical stage, stage clips will be absent.

Light Source: Built-in light to illuminate specimen on the stage. Some microscopes are not equipped with a built-in light source and a mirror. An external light source such as a desk lamp is placed in front of the mirror to direct the light into the microscope. The mirror has a concave side for sunlight and a flat side for artificial light such as a lamp.

Control (On/Off) Knob for Light Source: Knob on side of light source box.

Light Diaphragm: Variable shutter or disk containing small holes of various sizes; controls amount of light passing from light source.

Body Tube: Joins the nosepiece to the ocular lens.

Ocular Lens (Eyepiece): Lens through which you look into the microscope (magnification of the ocular lens is commonly 10X). Your microscope may be monocular or binocular.

Nosepiece: Rotating disk to which objective lenses are attached.

Objective Lenses: Lenses closest to specimen, each of which has a different magnification: 4X (scanning power), 10X (low power), and 40–45X (high power). While often not included in basic student microscopes, your microscope may also have an oil immersion lens (97–100X).

Coarse Adjustment Knob: Large knob near the nosepiece, used to make large adjustments in focus. NOTE: The coarse adjustment should never be used when the high power objective is in place!

Fine Adjustment Knob: Small knob near base, used to make fine adjustments in focus.

MICROSCOPE EXERCISE

1. Rotate the **nosepiece** until the scanning power **objective** clicks into place.

2. Turn on the **light source.** The **field of view** (area seen through the lens) should be illuminated.

3. Get a letter "e" or "system" slide and place it on the stage with the coverslip on top and the letter right-side-up so it can be read with the naked eye.

    Using the coarse adjustment knob, lower the objective to about 1 cm from the slide.

    Look through the **ocular;** focus by raising the body tube with the **coarse adjustment knob.** If you can't find the material on the slide, move the slide back and forth while looking through the microscope. (If you *still* can't see anything, ask your instructor for help!) Now, swing the low power objective into place. *Never* clean the lenses of a microscope with anything but specially treated lens paper.

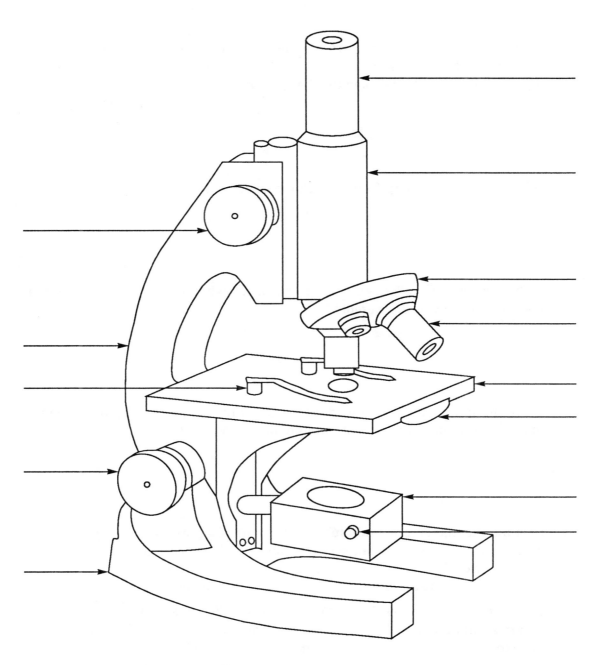

Figure 1-1. Parts of the compound microscope.

Look through the **ocular** and try the fine adjustment knob. Do you notice any difference? The fine adjustment is particularly useful at high power. Try adjusting the light intensity by using the **diaphragm.**

When the "e" is in sharp focus, sketch the letter "e" below.

Is the position of the letter the same as it looks when not viewed through the microscope? How is it different?

Move the slide to the left. How does this affect the apparent direction of movement of the slide as you view it through the microscope?

4. Keeping the microscope **body tube** in the same place and the slide in the same position, *slowly* swing the **high power objective** into place. It should just clear the surface of the slide. Microscopes are normally **parfocal;** once focused, use of other objective lenses should require only small adjustments in focus.

    CAUTION: *DO NOT* USE THE COARSE ADJUSTMENT KNOB ON HIGH POWER! You risk ramming the objective all the way down into the slide, possibly cracking expensive prepared specimens and damaging the objective.

5. Get a colored thread slide. Focus on scanning power. Then move the coarse adjustment knob slowly. Change to low power (10X) and fine focus. What do you notice about the material on the slide?

The technique of making small adjustments in focus to view different parts of a microscopic specimen is called **optical dissection** and will be particularly useful when you are viewing live material.

Draw your three threads; mark top, middle, and bottom.

Switch to high power. Follow the same technique on high power, using fine adjustment.

6. To estimate size of objects under the microscope, one must first determine the size of the **field of view.** Get a section of plastic ruler and place it over the opening in the center of the stage so that half the opening is covered. Focus on the metric ruler using scanning power.

7. Line up the vertical lines on the ruler so that one line is just visible at the left side of the field. Approximately what is the width of the field in mm?

For most microscopy, we use micrometers (μm), which equal one thousandth of a mm (0.001), or one millionth of a m (0.000001). What is the diameter of the low power field in (μm)?

Measure the field of view using your scanner, low, and high power objectives and then fill in this table:

Total Magnification	Field of View (mm)	(μm)
40X	_____	_____
100X	_____	_____
430X	_____	_____

8. As time permits, get materials from your instructor and use optical dissection and field of view to draw and measure objects under the microscope.

NOTE: Professional biologists usually learn to keep both eyes open while looking through the microscope. If you have a monocular microscope, this may seem uncomfortable and distracting at first, but it prevents eyestrain when doing microscopy for long periods.

At the close of the lab, click the lowest power objective into place, and slowly lower the nosepiece as far as it will go. Turn off the light source and wrap the cord loosely around the base. Using the proper carrying technique, return the microscope to its place in the cabinet, taking care not to bump the nosepiece on the shelf.

2   Biological Classification

EQUIPMENT NEEDED

None

TAXONOMY AND BINOMIAL NOMENCLATURE

There are literally millions of different kinds of living organisms on earth, many of which remain unnamed and undescribed. Biologists thus constantly face the problem of systematically naming and classifying new organisms. An efficient, universally accepted system of classification, or **taxonomy,** is essential if scientists are to identify, investigate, and exchange information about organisms.

In the 1700s, a Swedish naturalist named Carl von Linné, or "Linnaeus," originated **binomial nomenclature,** the naming system used by all biologists today. Under this system, every organism has a Latinized scientific name consisting of two parts, the **genus** and the **species.** Regardless of how many common names an organism may have in different languages and different parts of the world, this scientific name is generally agreed upon and officially registered within the scientific community. The scientific name for humans, for example, is *Homo sapiens,* or "wise man." (Note that the genus, *Homo,* is capitalized, whereas the species name, *sapiens,* is not. Scientific names are always set off with underlining or italics.)

Here are some examples of both common and scientific names for several organisms:

Common name	Scientific name
Human	*Homo sapiens*
Timber Wolf	*Canis lupus*
American Lobster	*Homarus americanus*
Sugar Maple	*Acer saccharum*
Apricot	*Prunus armeniaca*
Gray Squirrel	*Sciurius carolinensis*
Dog	*Canis familiaris*

The scientific name alone does not reveal much information about the relationships among species. Today, biologists use a seven-level hierarchical system of classification that attempts to depict phylogenetic relationships among all species. Lower levels are placed within each of the higher levels. The seven levels are:

Kingdom
    Phylum
        Class
            Order
                Family
                    Genus
                        Species

The following are examples of complete seven-level classifications:

	Human	Wolf	Dog	Fox Squirrel	Sugar Maple
Kingdom	Animalia	Animalia	Animalia	Animalia	Plantae
Phylum	Chordata	Chordata	Chordata	Chordata	Tracheophyta
Class	Mammalia	Mammalia	Mammalia	Mammalia	Angiospermae
Order	Primates	Carnivora	Carnivora	Rodentia	Sopindales
Family	Hominidae	Canidae	Canidae	Sciuridae	Aceraceae
Genus	*Homo*	*Canis*	*Canis*	*Sciurius*	*Acer*
Species	*sapiens*	*lupus*	*familiaris*	*niger*	*saccharum*

THE KINGDOMS

In the past, biologists recognized only two kingdoms of organisms: Plantae (plants) and Animalia (animals). However, not all organisms fit neatly into one of these categories. Some "plants" have a number of animal-like characteristics and vice-versa.

Today biologists generally recognize five kingdoms of organisms:

Kingdom MONERA: Primitive microscopic, unicellular (single-celled) organisms with no well-defined nuclear envelope or membrane-bound organelles (Prokaryotes); includes bacteria and blue-green algae. Blue-green algae are now commonly referred to as bacteria (Cyanobacteria). Some bacteria are autotrophs and some are heterotrophs.

Kingdom PROTISTA: Typically unicellular organisms with membranous organelles, most notably the nucleus. Includes autotrophs and heterotrophs. Representatives are the plantlike unicellular algae and the animal-like protozoa.

Kingdom FUNGI: Multicellular heterotrophs that typically absorb nutrients. Includes mushrooms, molds, and yeasts. All are eukaryotic and characteristically lack chlorophyll. Fungal cell walls contain chitin, a compound that is structurally very similar to cellulose.

Kingdom PLANTAE: Multicellular, eukaryotic autotrophs. Includes the multicellular algae and terrestrial plants. Contain chlorophyll and thus produce their own organic nutrients through photosynthesis. Most plant cells are enclosed in a rigid cell wall. Cellulose is the basic strengthening component.

Kingdom ANIMALIA: Multicellular, eukaryotic heterotrophs, including herbivores, carnivores, and omnivores. Numerous animals display motility and lack the rigid cell walls characteristic of plants. Includes jellyfish, worms, insects, fish, frogs, birds, mammals, and a great variety of other organisms.

Complete this table:

Organism	Description	Kingdom
Haliaectus leucocephalus	Carnivorous multicellular organism that feeds chiefly on dead or dying fish.	_____
Crassostrea virginica	Multicellular organism, in a two-part shell, that feeds by filtering tiny particles from water.	_____
Staphylococcus aureus	A bacterium that is commonly found on the skin and can produce infections.	_____
Amanita muscaria	A complex multicellular organism common in moist areas. Absorbs nutrients from its surroundings. Lacks chlorophyll. Chitin predominant in cell walls.	_____
Lycopersicon esculentum	Photosynthetic, multicellular organism with roots, stems, leaves, and fruit.	_____
Paramecium caudatum	Aquatic unicellular, eukaryotic, heterotrophic organism. Utilizes pseudopodia in movement.	_____
Rhizobium	A unicellular, prokaryotic organism that lacks a nuclear envelope. Capable of nitrogen fixation and lives in legumes.	_____
Procambarus clarkii	A multicellular, heterotrophic organism. Frequently caught in sufficient numbers to be marketed for food. Has numerous appendages, obvious body segmentation, and an exoskeleton.	_____
Saccharomyces cerevisiae	A single-celled eukaryotic organism. Chitin present in cell wall. Does not carry on photosynthesis. Absorbs nutrients from its surroundings. Used in bakeries and breweries.	_____

3 The Plankton Community

EQUIPMENT NEEDED

1. Compound microscopes
2. Prepared slides: *Paramecium caudatum* and *Spirogyra* sp.
3. Microscope slides and coverslips
4. Living cultures of *Paramecium* and *Spirogyra* sp.
5. Methyl cellulose (1.5% liquid) or other mobility inhibitors
6. Yeast powder stained with Congo Red dye
7. Toothpicks
8. Dilute acetic acid (1%)

The **plankton** community is an assemblage of aquatic microorganisms that swim weakly or float in lakes, rivers, and the oceans. This community includes representatives of several kingdoms, including monerans, protists, plants, and animals. Plantlike plankton are referred to as **phytoplankton;** animal-like plankton, as **zooplankton.**

Plankton are present in a great variety of ecosystems and provide a food source for many aquatic systems. A highly simplified example of this type of food chain is shown in Figure 3-1. Figure 3-2 shows a more realistic example. Planktonic organisms are a frequent subject of studies of water quality and the factors controlling productivity of aquatic ecosystems. Because they are extremely sensitive to water pollutants, they are used by environmental biologists as indicators of water quality.

LABORATORY EXERCISE I: PLANKTON REPRESENTATIVES

The two organisms you will examine in this laboratory, *Paramecium* and *Spirogyra,* can easily be found in local ponds and belong to the Kingdoms Protista and Plantae, respectively. *Spirogyra* is plantlike, whereas *Paramecium* is animal-like.

Paramecium (Fig. 3-3) is a protozoan common in freshwater ecosystems containing decaying vegetation. It feeds on bacteria, small protozoans, algae, and yeasts. *Paramecium* is **ciliated;** it uses the small hairlike organelles, or **cilia,** to propel itself through the water. In addition to its ability to swim and the fact that it is a heterotroph, *Paramecium* has a number of other animal-like characteristics. It lacks a rigid cell wall; it is capable of simple behaviors such as moving away from irritating or unfavorable stimuli such as light or heat. *Paramecium* may respond to unfavorable environmental conditions by discharging its **trichocysts,** or organelles of defense.

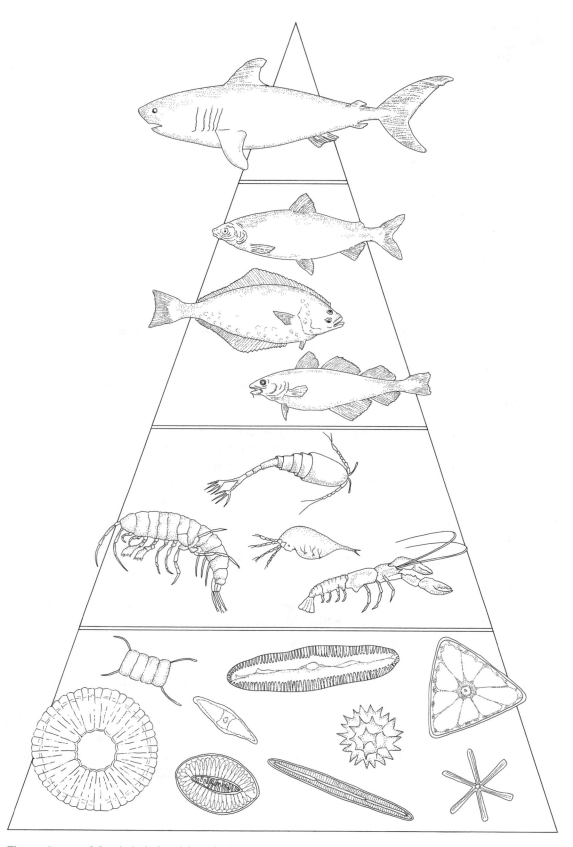

Figure 3–1. A food chain involving plankton.

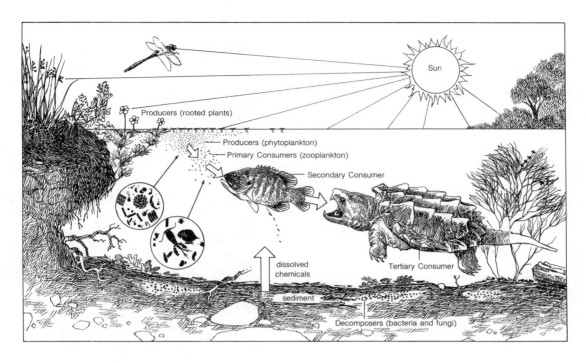

Figure 3–2. Plankton in the food chain in a freshwater pond ecosystem. (From G. Tyler Miller, Jr., *Living in the Environment,* 7th ed. Belmont, CA: Wadsworth, 1992, p. 88.)

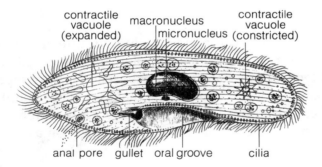

Figure 3–3. The structure of *Paramecium caudatum*, a freshwater ciliate.

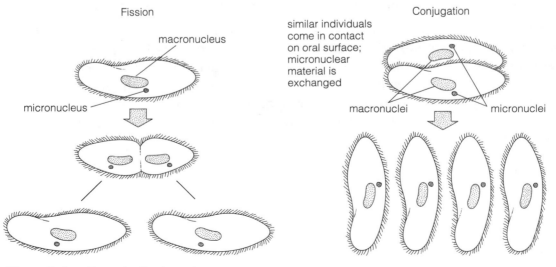

Figure 3–4. Reproduction in protozoans.

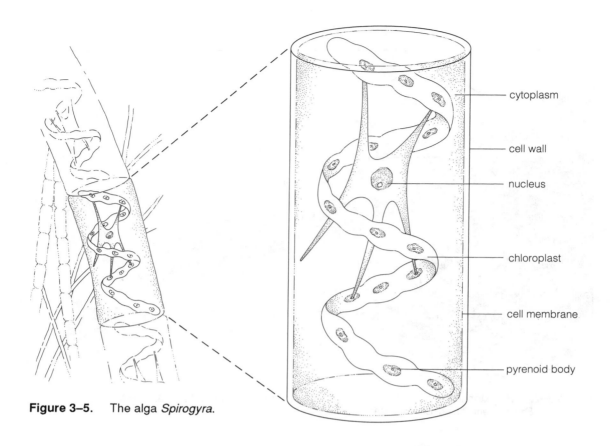

Figure 3–5. The alga *Spirogyra*.

cytoplasm

cell wall

nucleus

chloroplast

cell membrane

pyrenoid body

Paramecium, like other protozoans, can reproduce in two different ways (see Fig. 3-4). In the first, **fission,** a form of asexual reproduction, the parent cell divides, producing two genetically identical daughter cells. In the second type of reproduction, **conjugation,** two parent cells exchange genetic material via a **cytoplasmic bridge** formed between them, resulting in offspring genetically different from either parent. This exchange of genetic material qualifies conjugation as a form of **sexual reproduction.**

Spirogyra (Fig. 3-5) is a green alga commonly found floating in freshwater ponds, where it often forms light green masses on the surface. This particular algal species, commonly found in polluted waters, is often referred to as "pond scum."

Spirogyra is an autotroph and has a number of plantlike characteristics, which make it very different from *Paramecium*. *Spirogyra* has a rigid **cell wall** and **chloroplasts** (organelles of photosynthesis), and it is incapable of locomotion or behavior. Like *Paramecium*, *Spirogyra* also can reproduce in two ways. Fission will result in two genetically identical daughters, whereas conjugation involves the exchange of genetic material and the production of new individuals genetically different from either parent.

A. Prepared Slides

1. Obtain prepared slides of *Paramecium* and *Spirogyra*. The organisms on these slides have been killed, fixed (preserved), and stained to highlight their structural components.

2. Using the figures in this manual, locate and identify as many of the internal structures of each species as possible. In particular, look for evidence of reproduction, either sexual or asexual, on the prepared slides.

B. Wet Mounts of Living Material

These prepared slides are of living *Paramecium* and *Spirogyra*. Your instructor will demonstrate the proper procedure for preparing a "wet mount," which is as follows:

Place a small drop of liquid from the flask containing living *Paramecium* or *Spirogyra* on a clean glass slide. Rather than dropping a coverslip directly onto the sample, place one edge of the coverslip at an angle so that one edge just touches your sample, then gently lower the coverslip until the sample is completely covered. Be careful not to trap air bubbles underneath.

Examine your samples under the microscope, and attempt to locate as many subcellular structures of each species as possible. Take care not to touch the objective lenses to the wet mount. You may want to save any particularly good mounts of *Paramecium* for the following exercises.

C. Behavior and Physiology of *Paramecium*

After familiarizing yourself with the anatomy of *Paramecium*, and, working with either your original wet mount or a new sample, perform the following exercises:

1. Locomotion: The cilia of *Paramecium* beat backward, carrying the organism forward in the water. As the cilia stroke obliquely (at an angle), the *Paramecium* rotates on its longitudinal axis, moving ahead on a spiral course. To swim backward, the cilia beat in the reverse direction. Observe the normal behavior and locomotion of *Paramecium* as it responds to stimuli in its environment. Do you see evidence of either avoidance or attraction to objects or other organisms?

   Your instructor will show you how to use either a viscous liquid (methyl cellulose) or other mobility inhibitors to slow *Paramecium* down, making observation easier. Using high power, adjust the light diaphragm on your microscope until you can see the cilia beating.

2. Feeding: Using a toothpick, place a small amount of Congo Red dye in yeast on a slide containing a wet mount of *Paramecium*. Attempt to observe feeding behavior and the color change in the dye as food enters the food vacuole to be digested (Congo Red turns blue when the pH is sufficiently acid—blue below pH 3 and red above pH 5).

3. Excretion: Using nonmobile *Paramecium*, observe the action of the contractile vacuoles with their radiating canals. The rate of vacuole discharge is higher in water with a scant supply of dissolved salts than in water with higher concentrations of salts. Why?

4. Trichocyst Discharge: Following your instructor's directions, add a drop of dilute acetic acid to the edge of your coverslip. This stimulus will usually promote a discharge of *Paramecium's* trichocysts, which are elongated capsules located along the periphery of the *Paramecium*. The capsules discharge long, thin filaments which can be triggered by a variety of chemical or mechanical stimuli. The filaments are commonly ejected when a *Paramecium* is attacked by a predator. NOTE: The acid will eventually kill the organism, so save this activity for last.

D. Complete the following chart:

Characteristic	Spirogyra	Paramecium
Presence of cell wall?		
Function of cell wall		
Presence of cell membrane?		
Function of cell membrane		
Method of maintaining water balance		
Source of energy/food		
Method of locomotion		
Method of reproduction		

E. What do you see as advantages and disadvantages for asexual and sexual reproduction?

LABORATORY EXERCISE II: PLANKTONIC DIVERSITY

Community diversity, or the number of different kinds of organisms present in a given community, is thought by many ecologists to be an important indicator of environmental quality. There is thought to be a relationship between community diversity or complexity and **stability,** the ability of the community to recover after an environmental insult, such as pollution.

Many factors can influence the diversity of aquatic communities, including nutrient levels, depth, temperature, successional stage, and pollutants. In this lab you will compare the diversity of plankton in samples of several aquatic communities and relate these to the level of pollution.

Procedure

1. Prepare wet mounts from water samples as follows:
   a. Place a small drop of water from a sample (after shaking) on the microscope slide.
   b. Carefully place the coverslip over the drop.

2. Using Figure 3-6 as a guide, identify as many of the common plankton representatives in your sample as possible, keeping a list of the different kinds you see as well as the number of each type. Complete the chart on page 19.

3. Which aquatic community (collecting site) is the most diverse? The least diverse? Can you explain the differences in diversity in the samples you examined, based on differences in the abiotic characteristics of the ecosystems from which they were collected?

4. Which organisms are most numerous in samples from low-diversity communities? How might you account for their dominance (greater numbers)?

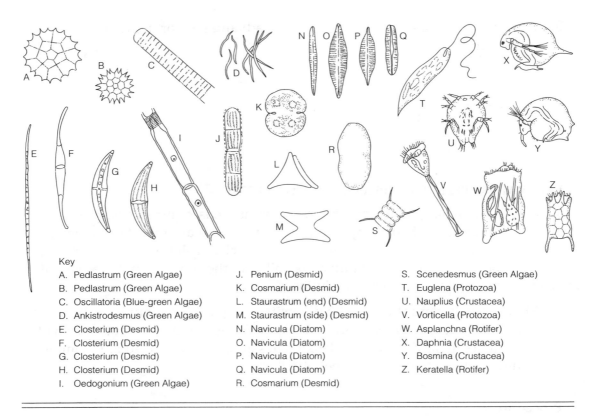

Key

A. Pedlastrum (Green Algae)
B. Pedlastrum (Green Algae)
C. Oscillatoria (Blue-green Algae)
D. Ankistrodesmus (Green Algae)
E. Closterium (Desmid)
F. Closterium (Desmid)
G. Closterium (Desmid)
H. Closterium (Desmid)
I. Oedogonium (Green Algae)

J. Penium (Desmid)
K. Cosmarium (Desmid)
L. Staurastrum (end) (Desmid)
M. Staurastrum (side) (Desmid)
N. Navicula (Diatom)
O. Navicula (Diatom)
P. Navicula (Diatom)
Q. Navicula (Diatom)
R. Cosmarium (Desmid)

S. Scenedesmus (Green Algae)
T. Euglena (Protozoa)
U. Nauplius (Crustacea)
V. Vorticella (Protozoa)
W. Asplanchna (Rotifer)
X. Daphnia (Crustacea)
Y. Bosmina (Crustacea)
Z. Keratella (Rotifer)

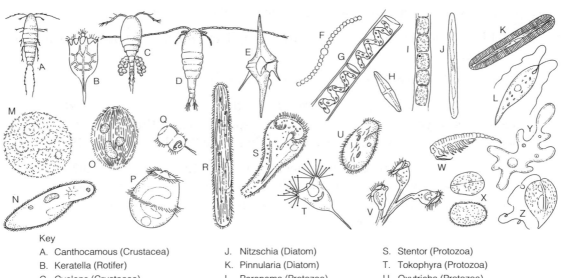

Key

A. Canthocamous (Crustacea)
B. Keratella (Rotifer)
C. Cyclops (Crustacea)
D. Diaptomus (Crustacea)
E. Ceratium (Protozoa)
F. Anabaena (Blue-green Algae)
G. Spirogyra (Green Algae)
H. Stauroneis (Diatom)
I. Microspora (Green Algae)

J. Nitzschia (Diatom)
K. Pinnularia (Diatom)
L. Paranema (Protozoa)
M. Volvox (Protozoa)
N. Paramecium (Protozoa)
O. Prorodon (Protozoa)
P. Didinium (Protozoa)
Q. Urocentrum (Protozoa)
R. Spirostomum (Protozoa)

S. Stentor (Protozoa)
T. Tokophyra (Protozoa)
U. Oxytricha (Protozoa)
V. Carchesium (Protozoa)
W. Eubranchipus (Crustacea)
X. Cypridopsis (Crustacea)
Y. Amoeba (Protozoa)
Z. Phacus (Protozoa)

Figure 3–6. Some common freshwater plankton.

Collecting Site		Collecting Site		Collecting Site		Collecting Site	
Plankton Type	Number of Individuals	Plankton Type	Number of Individuals	Plankton Type	Number of Individuals	Plankton Type	Number of Individuals

4 Ecological Succession

EQUIPMENT NEEDED

1. Field guides to plants and animals are useful.
2. 35mm slide series on Primary and Secondary succession (useful for inclement weather). Slides may be made by instructor or purchased from numerous supply houses.

In ecology, a **community** is defined as an aggregate of organisms that forms a distinct ecological unit. Rather than being stagnant, communities are constantly changing, as are the individual populations of organisms that form parts of the larger, more complex whole. This process of ecological change, or gradual replacement of one community by another, is known as **succession.** Succession occurs in aquatic as well as terrestrial ecosystems.

Succession may occur slowly and take hundreds or thousands of years to reach the successional stage we now see. Communities that have been disturbed or destroyed almost overnight by human activities may recover quickly or take hundreds of years to recover, if they are able to recover at all.

PRIMARY VS. SECONDARY SUCCESSION

Many aspects of community evolution are not well understood, and the topic of ecological succession is a controversial one among ecologists. In general, however, ecologists agree that certain types of communities undergo a predictable sequence of changes. Each stage, or **seral stage**, in the successional process is characterized by certain dominant forms of plants and animals, which are eventually replaced in the next seral stage. It is believed that many community types ultimately reach a stable stage in which no further change occurs. This final seral stage is known as the **climax community.**

Two types of predictable successional sequences have been described for terrestrial communities. **Primary succession** is defined as succession in areas that are devoid of living organisms—for example, bare rock, volcanic deposits, and sand dunes. (Sand is the product of mechanical pulverization of various rocks.) **Secondary succession** (see Fig. 4-1) occurs after community change following disturbance by human activities or natural events. Since soil and some organisms are still present, secondary succession occurs more rapidly than primary succession. However, even "old field succession," the reforestation of abandoned farmland, can still take hundreds of years.

canopy

lower canopy trees

tall shrub understory

annual weeds

perennial weeds and grasses

shrubs

young pine forest

mature oak forest

Time ──────▶

Figure 4–1. Secondary succession of plant communities in an abandoned farm field in North Carolina. Time scale is about 150 years. Succession of animal communities is not shown. (From G. Tyler Miller, Jr., *Living in the Environment*, 7th ed. Belmont, CA: Wadsworth, 1992, p. 158.)

SUCCESSION IN FRESHWATER AQUATIC ECOSYSTEMS

Aquatic communities, as well as terrestrial ecosystems, undergo change. Most freshwater lakes and ponds were formed by glaciation or by other events that occurred relatively "recently" in geologic terms. For example, the Great Lakes were probably formed about 10,000 years ago. In aquatic succession, lakes gradually become richer in organic material and in concentrations of inorganic nutrients such as nitrogen and phosphorus. This process is referred to as **eutrophication** (see Fig. 4-2). As organic material accumulates in the lake bottom and aquatic plants begin to spread from the edges to the middle of the lake, the lake becomes progressively shallower. Ecologists generally agree that freshwater lake communities, including the Great Lakes, will eventually become filled in, passing through shallower and shallower seres. In the latter stages of eutrophication, lakes become swamps or marshes before ultimately becoming terrestrial rather than aquatic communities.

The rate of eutrophication depends on a number of factors, including initial depth of the lake or pond and the amounts of nutrients available from the surrounding ecosystem. Ecologists rate lakes according to their successional stage on a continuum from **oligotrophic,** or relatively nutrient-poor, to **eutrophic,** or highly enriched. Water pollution caused by human activity can hasten or otherwise disrupt the normal, gradual eutrophication process, essentially "aging" freshwater ecosystems prematurely.

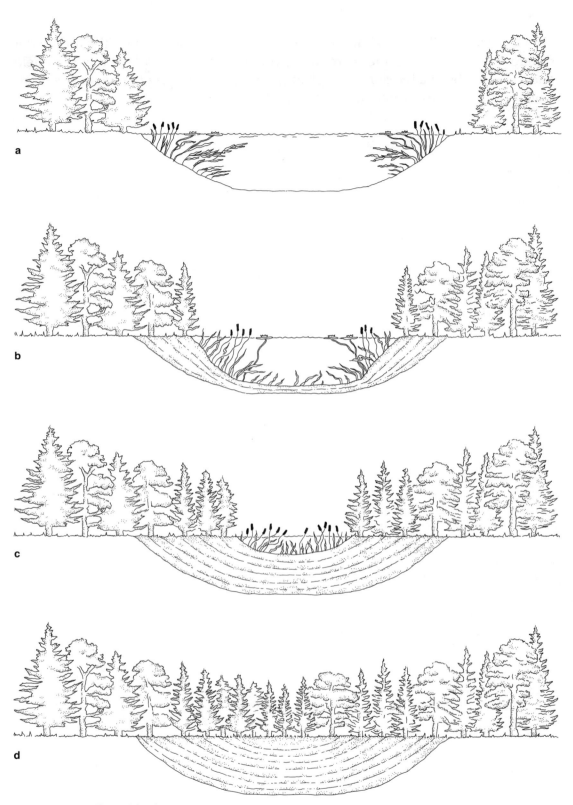

Figure 4–2. Eutrophication.

LABORATORY EXERCISE

Your instructor will show slides on succession in different communities and under different conditions. Weather permitting, you may take a short hike to a local field and pond to observe the dominant plant species characterizing the stages of succession, or seres, in freshwater and field ecosystems.

Exponential Growth

EQUIPMENT NEEDED

None

Once upon a time, a king who had won a chess match with another king told the loser he wanted some wheat for his prize. The amount of wheat was to be determined by counting out the grains using the squares on a chess board. The first square would represent one grain of wheat, the second would represent two, the third four, and so forth, with each square being worth double the previous one and continuing through all 64 squares.

Figure 5-1 shows the effect of such doubling, and Table 5-1 shows the numbers that result when doubling 32 times. (The first doubling, of 1, produces 2, etc.) Remember that the doublings shown would complete only half the chess board. The numbers would be very much larger by the time the end of the chess board was reached. A growth increase such as this is referred to as **exponential growth.**

Exponential growth can occur with populations of organisms, resource consumption, generation of waste, and energy demand. The human population also shows it. We might expect energy and resource use to increase as the population increases, but if people use more energy or resources as time goes on, the demand for resources will, of course, increase at a greater rate than the population.

Notice the "snowball" effect that exponential growth has. Step 1 adds 1 unit to the previous amount, step 2 adds 2, step 3 adds 4, step 4 adds 8, and so on. By the time we get down to step 22, millions of units are being added. Doubling a small number produces a small change. Doubling a big number produces a big change, so the increase in whatever we are measuring accelerates as time passes, and the curve turns more steeply upward.

Now, think about putting marbles in a jar (to represent population growth) or taking them out (to represent use of nonrenewable resources). You add or subtract the marbles at one-minute intervals, and do so according to Table 5-1. Once the jar is half full (or half empty), it takes only one more minute for it to be completely full (or empty).

0	1		17	131,072
1	2		18	262,144
2	4		19	524,288
3	8		20	1,048,576
4	16		21	2,097,152
5	32		22	4,194,304
6	64		23	8,388,608
7	128		24	16,777,216
8	256		25	33,554,432
9	512		26	67,108,864
10	1,024		27	134,217,728
11	2,048		28	268,435,456
12	4,096		29	536,870,912
13	8,192		30	1,073,741,824
14	16,384		31	2,147,483,648
15	32,768		32	4,294,967,296
16	65,536		33	8,589,934,592

Table 5–1. Results of doubling.

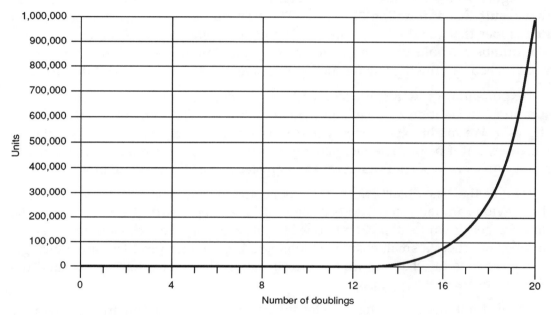

Figure 5–1. Exponential growth.

Instead of using the type of graph you are used to, like the one in Figure 5-1, scientists sometimes use what is called a semilog grid, like Figure 5-2. The grid has time intervals, or steps, indicated across the bottom in familiar fashion, but the logarithmic spacing on the vertical axis is probably not familiar to you. In this case, you would read the lines above each numbered line as multiples of the first labeled line preceding, as shown in the two examples following:

Example 1	Example 2	Explanation
1000—	**1200—**	Labeled line
900—	1080—	Labeled line below x9
800—	960—	Labeled line below x8
700—	840—	Labeled line below x7
600—	720—	Labeled line below x6
500—	600—	Labeled line below x5
400—	480—	Labeled line below x4
300—	360—	Labeled line below x3
200—	240—	Labeled line below x2
100—	**120—**	Labeled line
90—	108—	Labeled line below x9
80—	96—	Labeled line below x8
70—	84—	Labeled line below x7
60—	72—	Labeled line below x6
50—	60—	Labeled line below x5
40—	48—	Labeled line below x4
30—	36—	Labeled line below x3
20—	24—	Labeled line below x2
10—	**12—**	Labeled line

EXERCISE

1. Carefully plot several points from the top, middle, and bottom of Table 5-1 on the grid in Figure 5-2. You will have to estimate where some points go.

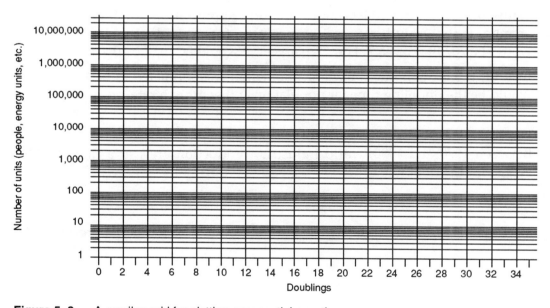

Figure 5–2. A semilog grid for plotting exponential growth.

2. Now fit a curve through the points as well as you can. What shape does the curve have?

3. You would need to extend the curve on an exponential growth graph to predict what values might occur in the future. On which type of grid would it be easiest to do that, an ordinary grid like Figure 5-1 or a semilog grid like Figure 5-2?

4. Refer to Table 5-1. If the steps each take 1 day, how long would it take to go from 1 unit to 2?

How long would it take to go from 524,288 units to 1,048,576 units?

Exponential growth results in a periodic doubling, as previously shown. If the growth rate is given as a percentage, the doubling time can be obtained by dividing that percentage into 70. This is called the Rule of 70, and it applies to compound interest at the bank or any other application involving a percent change. If the growth rate is 10% per year, for example, the doubling time would be 70/10 = 7 years. For 7% per year, it would be 70/7 = 10 years.

5. In 1991, the growth rate for world population was estimated to be 1.7%. What would be the doubling time for population at that growth rate?

6. Between 1970 and 1990, world industrial production grew at a rate of 3.3%. What was the doubling time?

7. a. List some resources that are used in connection with industrial production.

   b. Explain why some of these are unlikely to be available to sustain a 3.3% growth rate in industrial production.

A straight segment on a semilog graph indicates exponential growth, whereas the steepness of the segment indicates the growth rate. Examine the graph in Figure 5-3 for such trends.

8. a. Between which years did exponential growth occur for 60 years?
   _____ and _____

   b. When was there exponential growth for 30 years?
      Between _____ and _____.

   c. Which interval had the greatest growth rate?
      Between _____ and _____.

Obviously, exponential growth cannot go on indefinitely as far as population, resource use, or other things are concerned. There are limits. Either a leveling off or a decline has to occur eventually.

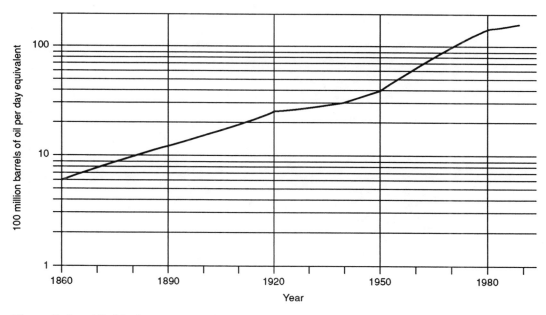

Figure 5–3. World primary energy demand.

INTERESTING NOTE

An important use of exponential growth is its application to the decay of radioactive materials. As a radioactive element (radioisotope) gives off radiation, it changes (decays) to a stable (nonradioactive) element. The rate at which this occurs is called the half-life of the radioisotope. The half-life is the length of time required for half of a quantity of radioisotope to decay. Half-lives range from as little as a fraction of a second to as much as millions of years.

Radioactive decay reduces the amount of radioisotope in a way that is exactly the opposite of the doubling discussed above. With each half-life, the amount of remaining radioisotope is cut in half, so after the first half-life has passed there is one-half the original amount of radioisotope, after the second half-life has passed there is one-fourth, after the third there is one-eighth, and so forth. This has implications for the storage of radioactive waste. With the passage of each half-life, the radiation is cut in half, because the radiation is related to the amount of radioisotope present. (Actually, radioactive waste often contains a variety of radioisotopes with different half-lives, and those with short half-lives decay rapidly, leaving those with long half-lives to linger on for many years.)

6   Population Control

EQUIPMENT NEEDED

1. Compound microscope
2. Prepared slides of ovaries from hamster or other small mammal

EXERCISE 1

Your instructor will conduct a discussion on possible methods to be used in stabilizing the human population. For this lab to be effective, class participation is very important. Before coming to class, you should be prepared to discuss the following topics:

1. Voluntary Control of Population Growth

   a. Do you feel that current birth control methods are adequate?

   b. Is voluntary abortion a suitable substitute for contraceptive use?

   c. What is an "ideal" contraceptive?

   d. Should unmarried teenagers have free access to birth control counseling and services without parental permission?

   e. What are the pros and cons of sterilization?

2. Involuntary Control of Population Growth

   a. What is the feasibility of limiting family size to one or two children?

   b. Should the size of a family be determined by its economic status?

   c. Concerning involuntary population control, what do you see as the role of compulsory sterilization or abortion for those couples exceeding the mandated allotment of children?

d. Can immigration limitations realistically play a role in population control?

e. Are you aware of any countries that currently utilize involuntary population control programs?

3. Legislative and Economic Support for Population Control

a. Should population control efforts constitute a major goal for the United States; for countries low on the socioeconomic scale; for the world?

b. How will future changes in our population structure affect the world economy?

c. Should the elimination of income tax deductions for dependents be used as a possible inducement for having smaller families?

d. Could population control efforts affect our Social Security system?

EXERCISE 2

The ovaries produce ova, discharge ova (ovulation), and secrete sex hormones. Although the ovaries play an essential role in the reproductive process, they are not very large. In humans, the paired ovaries resemble unshelled almonds in both size and shape. Using your compound scope, examine hamster ovaries or other small mammal ovaries and identify **immature follicles, maturing follicles, mature follicles** (Graafian follicle), and **corpora lutea** (Fig. 6-1). The **ovarian follicles** are eggs and their associated tissues in various stages of development. **Graafian follicles** are endocrine glands made up of a mature ovum and surrounding tissues. Graafian follicles not only contain a mature egg but also function as a hormone-secreting gland, secreting estrogens. The **corpus luteum** is also a hormone-producing, glandular body that develops from a Graafian follicle after ovulation. This gland secretes progesterone, estrogens, and relaxin. Note that the ovarian tissue you are examining is "sectioned": consequently, a mature follicle may have been cut in such a way that no ovum is present.

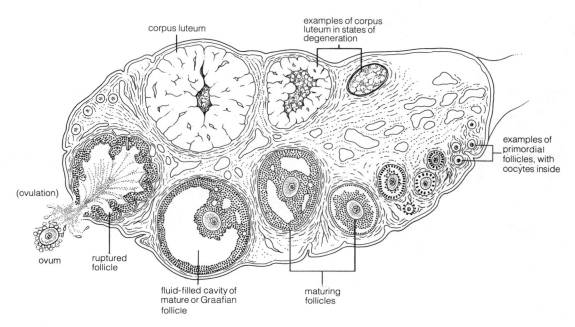

Figure 6–1. An ovary in cross section. (From Cecie Starr and Ralph Taggart, *Biology: The Unity and Diversity of Life*, 5th ed. Belmont, CA: Wadsworth, 1989, p. 514.)

The Bacteria: Representatives of Kingdom Monera

EQUIPMENT NEEDED

1. Nutrient agar petri plates
2. Sterile cotton swabs
3. Soil and some type of living insects are useful
4. Glass slides
5. Bunsen burner
6. Methylene blue stain in staining dish
7. Incubator (32°–35°C)
8. Compound microscopes preferably with oil immersion lenses and immersion oil
9. Safety goggles must be worn

Bacteria are microscopic, unicellular (one-celled) organisms belonging to the Kingdom Monera. Unlike the protists (eukaryotes), which have a well-defined nuclear envelope and other organelles, bacteria (prokaryotes) have no membrane separating their genetic material from the rest of the cytoplasm. It is thought that bacteria were among the earliest organisms to evolve, and they still retain many characteristics of primitive cells.

Despite their superficial simplicity, bacteria are probably adapted to a greater variety of environments than almost any other kind of organism. They are nearly ubiquitous (present everywhere) and can be found on land, in water, inside plants and animals, and even in environments too warm, too cold, or too polluted for other organisms to survive. Many species are **aerobic,** requiring oxygen to live and reproduce, whereas **anaerobic** bacteria thrive in environments lacking the oxygen essential for other forms of life.

Bacteria can exploit a number of different energy sources, and they directly or indirectly affect virtually all major functions of the biosphere. They play a variety of ecological roles. Most are **heterotrophs** and must consume other organisms or their products to survive. Others are **autotrophs;** like green plants (and many protists) they can manufacture their own sugars and other energy-containing molecules from inorganic materials.

Bacteria and other prokaryotes influence the well-being of humans in many ways, and they are extremely important to environmental health. Bacteria hasten the spoilage of food and, in some cases, can cause food poisoning. **Pathogenic,** or disease-causing, bacteria bring about certain illnesses in plants and animals, including tuberculosis, pneumonia, syphilis, and diphtheria. The vast majority of bacteria are harmless, however, and many are beneficial to humans. They are used in the commercial production of vinegar and of dairy products, such as cheese and yogurt In addition, they are used by industry in the manufacture of enzymes, antibiotics, hormones, insulin, and other life-saving chemicals.

Bacteria also occur as normal and occasionally beneficial inhabitants of the human intestinal tract. These intestinal bacteria are referred to collectively as the "intestinal flora." Elimination of normal flora by certain antibiotics often leads to overgrowth by resistant forms such as *Clostridium difficile* and many times results in disease (e.g., pseudomembranous colitis).

Perhaps most importantly, bacteria and other prokaryotes are abundant in healthy ecosystems of all types, and they constitute an essential component of all food chains. Without bacteria there would be no significant decomposition and no significant recycling of nutrients in ecosystems. One group, the **nitrogen-fixing bacteria,** plays a key role in the nitrogen cycle, converting nitrogen gas in the air to nitrogen-containing compounds that can be used by plants and, in turn, by other organisms, including humans.

BACTERIAL CLASSIFICATION AND MORPHOLOGY

The functional bacterial unit consists of a single prokaryotic cell. Bacterial cells are classified according to their shape, or **morphology.** There are three basic morphological types: **bacilli** (rod-shaped), **cocci** (sphere-shaped), and **spirilla** (corkscrews) (see Fig. 7-1). Although single bacterial cells are capable of functioning independently, the cells of most species reproduce rapidly and form simple colonies. Because bacterial colonies consist of thousands of cells, and because different species form slightly different types of colonies, bacteriologists can often identify bacterial types based on the appearance of the colonies.

Because bacterial species have such a wide range of nutritional and environmental requirements, bacteriologists can also use chemical tests to identify them. One way to do this is to provide bacteria with an artificial growth medium containing known concentrations of certain nutrients and chemicals. By varying the composition of the growth medium and examining the colonies that appear, scientists can often determine what types of bacteria are present.

EXPERIMENTAL PROCEDURES (WEEK 1)

1. Your instructor will divide your class into several groups. Each group will be given a plastic petri plate containing sterile nutrient agar, a generalized medium that will allow the growth of a variety of bacterial types.

2. Using the method indicated in Figure 7-2, inoculate your plate with one of the following: throat, skin, soil, flies, air, and sterile swab. Using a wax pencil, label your plate by material tested and your name. The last group will serve as an experimental control.

3. After inoculation, plates will be incubated at room temperature for 1 week. At that time you will be asked to compare all of the groups' plates, compare the growth of colonies, and identify bacterial types under the microscope.

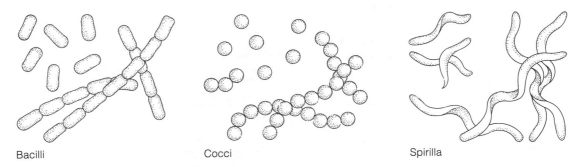

Bacilli   Cocci   Spirilla

Figure 7–1.   Bacterial morphology.

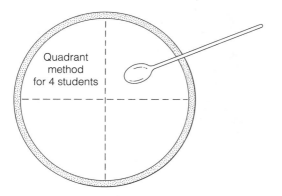

Quadrant
method
for 4 students

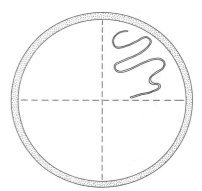

Figure 7–2.   Inoculation methods.

EXPERIMENTAL RESULTS (WEEK 2)

Source	Number of Colonies	Colony Types	Notes
Sterile swab (control)			
Air			
Throat			
Skin			
Soil			
Flies			

MICROSCOPIC EXAMINATION AND IDENTIFICATION OF BACTERIA

Before we view bacteria under the microscope they must be "fixed" (killed and stuck to a glass slide) and stained with special chemicals to make them more easily visible.

Procedure

1. Put a drop of water on a clean glass slide. With a wooden stick, place a small bit of one or two of the cultures into the water drop and mix thoroughly.

2. Smear the drop around on the slide to make a thin film.

3. Air-dry the slide and then pass it a few times—quickly!—through the flame of a Bunsen burner. The slide should feel warm, NOT HOT, to the touch.

4. Using a staining dish or jar, dip the slide in methylene blue stain for 30 to 60

seconds, then flood the slide gently with tap water on the back side. Dry the slide with blotting paper.

5. Examine your preparation under a microscope, preferably a scope with an oil immersion lens (97–100X).

    a. Get the bacteria on the slide in focus at low power.

    b. Slowly move the high power objective into place and adjust focus (fine focus only).

    c. Slowly move objectives so that the oil immersion lens is almost in place.

    d. Place a drop of immersion oil on the slide where the light shines through.

    e. Swing oil lens until it clicks in place.

    f. Adjust fine focus.

    g. Draw the bacteria present.

    h. After you're finished, wipe lens off with specially treated lens paper.

Discussion Questions

1. Which of the plates inoculated by the class contained the most bacteria overall, based on number and size of colonies?

2. Did the "control" plates grow any bacteria? What do you conclude about the experimental procedure from this result?

8 Water Quality Testing I: The Coliform Test

EQUIPMENT NEEDED

1. Select water samples
2. Lactose presumptive test kits (lauryl tryptose may also be used)
3. BGLB test kits
4. EMB plates
5. Incubator (32°–35°C)

One of the most important sources of water pollution is human fecal contamination. This is particularly true in undeveloped countries, where the drinking water supply is rarely adequately separated from water used for sanitation (i.e., for bathing, laundry, and toilets). Numerous diseases are spread via feces in drinking water, including typhoid fever, cholera, hepatitis, and dysentery. The **pathogens** (disease-causing microorganisms) are typically tasteless, odorless, and too small to be seen with the naked eye. They are very difficult to detect using routine laboratory procedures, and it would be very expensive to test for *all* types of potentially dangerous microorganisms, especially because many pathogens gain entrance to the water supply only sporadically. Pathogenic bacteria are typically very short-lived and present only in small numbers; hence they often escape detection by laboratory tests. Fortunately, fecal coliform testing is inexpensive and a highly effective alternative to testing for specific pathogenic species.

Coliforms are rod-shaped bacteria *(bacilli)* distinguished from most other species by the fact that they can promptly ferment **lactose**, or milk sugar, and produce gas within 48 hours at 35°C. There are two main types. **Fecal coliforms**— a primary representative is *Escherichia coli (E. coli* for short)— are found in human and animal intestinal tracts. Other coliform species, including *Enterobacter aerogenes*, do not exclusively inhabit the gut but also live in soil, grain, hay, and vegetable matter.

Coliform bacteria are *not* usually pathogenic; they are, in most cases, harmless. However, because the presence of either type of bacteria in the drinking water supply is evidence of contamination by human feces or soil, these organisms are used as *indicators* of water pollution. If *E. coli* is present, this is an indication of fecal contamination, which suggests the possibility that disease-causing bacteria could also be entering the water supply. If *E. aerogenes* is detected, it suggests contamination by soil or feces.

When fecal contamination has occurred, coliforms are present in large numbers. Unlike many pathogens, they are relatively long-lived and easy to detect. Coliform testing is a simple, reliable, and inexpensive procedure, performed routinely at sewage and water treatment plants. We will perform the first stage of testing, called the **presumptive coliform test.**

THE PRESUMPTIVE COLIFORM TEST

The presumptive coliform test makes use of the fact that all coliforms ferment lactose (milk sugar) for food and release carbon dioxide (CO_2) and hydrogen (H_2) gas in the process.

Each student group will be provided with a presumptive coliform test tube: a small vial with a screw cap containing a smaller, inverted tube (the Durham tube) inside (see Fig. 8-1a). *Do not open the test tube until your instructor tells you to do so.* The brown material in the tube is 2 mL of lactose media (concentrated). If you prepare the medium yourself, use 195 g per 1,000 mL of H_2O. If you prefer, you can purchase presumptive test tubes directly from supply houses such as the HACH Company.

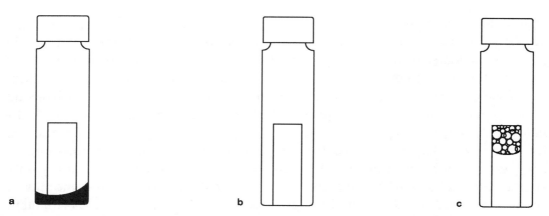

Figure 8–1.   Tube (containing inverted Durham tube) for use in the presumptive coliform test.

Procedure

1. Wash your hands thoroughly to reduce contamination.

2. Inoculate the tube with 10 mL of H_2O (inoculum) from one of several sources, as specified by your instructor. Tubes designed to contain 10 mL of sample when filled level-full are preferable.

3. Fill the tube with water from the appropriate sample. Screw the cap on tightly and label the tubes so that you can identify the sources of the water samples.

4. Shake the tube vigorously to dissolve the lactose and to completely detach the smaller Durham tube from the inside of the vial. When the lactose is completely dissolved and the Durham tube moves freely, turn the test tube upside down briefly to release all air bubbles. When the Durham tube is completely free of bubbles or foam (Fig. 8-1b), the tube is ready to incubate in the upright position (35°C for 48 hours).

Results

If lactose-fermenting bacteria such as coliforms were present in your water sample, gas will appear in the Durham tube (Fig. 8-1c). If no gas forms, no lactose fermenters were present; this constitutes a negative test result. Coliform testing is typically performed using a number of test tubes or replicates for each water sample being tested and the tubes are interpreted in sets of five or "five-tube assembly." Results of the presumptive coliform test are interpreted according to established criteria. If negative results (no gas bubbles) in all five tubes are obtained, the water is determined safe for drinking and testing is terminated. Fill in the following chart based on your data and data from other classes provided by the instructor.

Water Sample Tube	#1	#2	#3	#4	#5	Coliform #

Interpreting Test Results When Using Five-Tube Procedure*

No. Negative	No. Positive	Most Probable Number of Coliforms Per 100 mL
0	5	more than 16.0
1	4	16.0
2	3	9.2
3	2	5.1
4	1	2.2
5	0	acceptable for drinking water

*MPNs available from *Standard Methods*, 13th edition.

The presence of lactose-fermenting bacteria in presumptive coliform test vials however, is only a *preliminary* indication that coliforms are present; we presume at this point that the water being tested is contaminated and unsafe for human consumption. The presumptive test has the advantage of being cheap and easy to perform.

Not all lactose fermenters are coliform, however, and not all coliforms are the result of human fecal contamination. To verify the presence of coliforms and to precisely determine their source (feces or soil), testing must be carried several steps further (see Fig. 8-2).

CONFIRMED COLIFORM TESTING

Confirmed coliform testing is used to determine whether the bacteria producing positive presumptive test results are coliforms or some other lactose-fermenting species. The confirmed phase has two parts: the BGLB (Brilliant Green Lactose Bile) Tube and the EMB (Eosin Methylene Blue) Plate. In both kinds of test the bacterial cultures from positive presumptive test vials are used to inoculate new culture media. The use of BGLB tubes identical in size and shape to the presumptive tubes will simplify testing. You can inoculate the BGLB tubes by simply transferring the presumptive tube cap to the BGLB tubes. The EMB plates are simply streaked with the presumptive test inoculum.

The BGLB tube contains a bright green solution that inhibits growth of all lactose fermenters other than coliforms. After a suitable period of incubation, the formation of a gas bubble (CO_2) in the Durham tube of a BGLB vial confirms the presence of coliforms.

The EMB plate also selectively inhibits growth of bacteria other than coliforms. It has the added advantage that the special nutrient agar used in this test enables *E. coli* and *Enterobacter aerogenes* to form characteristically colored colonies: *E. aerogenes* produces pink or purple mucoid colonies, and colonies of *E. coli* have a green metallic appearance. You will have an opportunity to look at and interpret BGLB and EMB results in lab.

A positive confirmed test for coliforms indicates that the water supply is contaminated with feces, soil, or both. This suggests a dangerous opportunity for pathogenic microorganisms to transmit diseases among members of the population who are using this water for drinking, cooking, or bathing. Steps should be taken to correct this situation.

Positive results of a presumptive test would be followed in a water treatment plant by a confirmed test. The general scheme for coliform testing is outlined on the flow chart in Figure 8-2.

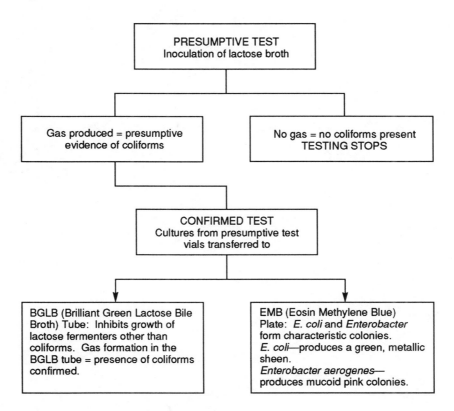

Figure 8–2. Flow chart of coliform testing procedures.

Water Quality Testing II: Dissolved Oxygen and Biochemical Oxygen Demand

EQUIPMENT NEEDED

1. Dissolved oxygen test kit (authors use Hach® D.O. test kit—HACH Chemical Co.)
2. B.O.D. respirometer (authors use Hach® B.O.D. respirometers)
3. Applicable water samples

Most living organisms, including aquatic organisms, require certain levels of oxygen to carry out normal metabolic processes. They are thus "aerobic" organisms (oxygen-dependent). The D.O. (dissolved oxygen) of "healthy" ecosystems typically ranges from about 4 to 8 ppm (mg/liter). In general, a D.O. of below 4 ppm (mg/liter) in a river or lake represents a very unhealthy situation for fish and other organisms.

The maximum amount of D.O. that a given aquatic ecosystem can hold depends on atmospheric pressure and water temperature; cold water retains more dissolved oxygen. However, the amount of D.O. in the system can also depend on the amount of organic material present.

One of the greatest challenges to the health of an aquatic community is the addition of large amounts of organic matter, such as sewage, garbage, or plant and animal wastes. Although these pollutants are highly **biodegradable** (capable of being broken down by normal biological processes), a healthy aquatic ecosystem can handle only so much of them before it becomes overloaded. Organic material is oxygen-demanding waste, which means that decomposer bacteria require oxygen to break it down. When a body of water becomes overloaded with **oxygen-demanding waste,** oxygen-using bacteria can deplete the D.O. content of the water below the level needed to support the diversity of organisms characteristic of healthy ecosystems.

The B.O.D. (biochemical oxygen demand) is a measure of the amount of organic waste in an aquatic ecosystem. Precisely defined, the B.O.D. is the amount of oxygen (mg/liter) consumed by bacteria in a sample of water kept at 20°C (68°F; about room temperature) over a 5-day period. Raw (untreated) sewage usually has a B.O.D. of 100 to 200.

There is an inverse relationship between D.O. and B.O.D.: as biochemical oxygen demand (organic content) goes up, D.O. goes down, because oxygen is consumed during the decomposition process. Both parameters are important measures of water quality.

Besides increasing B.O.D. directly, organic wastes can also indirectly raise the demand for oxygen. Organic wastes generally contain high concentrations of nitrogen and phosphorus, substances that act as limiting nutrients for plants. Because nitrogen and phosphorus are usually present in ecosystems only in very small concentrations, they act as "fertilizers" when added to lakes and streams in larger amounts. Nitrogen and phosphorus pollution can cause unnatural blooms of algae and other aquatic plants. When these plants die and begin to decay, aquatic microorganisms consume large amounts of oxygen in the decomposition process. The resulting abrupt decrease in D.O. is called an "oxygen crash," and can, in turn, bring about massive fish die-offs. Ultimately, this kind of pollution can reduce a healthy ecosystem to a smelly, virtually lifeless sewer inhabited by select bacteria or other organisms suited to **anaerobic** (no-oxygen) conditions. You should remember that most aquatic organisms are **aerobic** (oxygen-dependent).

THE D.O. TEST

Several water samples from local ponds or streams will be provided, and your lab table group will be assigned a particular water sample. Your objective, as a class, will be to measure and compare the amount of dissolved oxygen (D.O.) in the different samples using a standardized test kit.

The Hach® Dissolved Oxygen Test is a simplified procedure that makes use of premeasured reagents (chemicals of known concentrations). When there is oxygen in the water sample, the test solution turns bright yellow. The precise D.O. content can then be determined by measuring the amount of phenyl arsine oxide (a "titrant") needed to restore the sample to its original transparent state. Detailed, step-by-step instructions will be included in the Hach® kit, and your instructor will demonstrate the procedure to the class. Your lab instructor may elect to use a D.O. test kit other than Hach®.

> NOTE: It can be difficult to stopper the D.O. bottle without trapping one or more air bubbles inside. To prevent this, incline the D.O. bottle slightly and insert the stopper with a quick push. This should force any air bubbles out. However, if bubbles are trapped in the D.O. bottle in Step 2, the sample should be considered no longer useful and discarded. You should then start the test again.

After using the Hach® test to determine the D.O. content of your sample, the next step is to determine the **percent oxygen saturation** of the sample. In other words, you will determine what proportion of the oxygen that could potentially be held by water at the temperature of your sample is actually dissolved in the sample. To figure percent saturation, you need to know the D.O. and temperature of the water sample. You then use a special diagram (Fig. 9-1) called **Rawson's nomogram** to find percent saturation. Place dots on the diagram to mark the temperature and D.O. of a sample and, using a ruler, draw a line and connect the two dots. The point at which this line crosses the "% Saturation" line gives you the percent saturation of your sample.

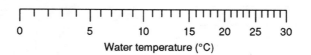

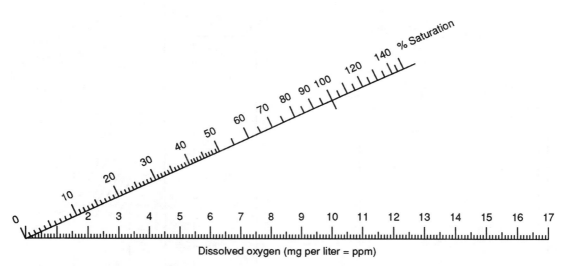

Figure 9–1. Rawson's nomogram. (Redrawn from Rawson, Special Publication No. 15, *Limn. Soc. Amer.*, 1944.)

Be prepared to discuss the following questions:

1. Which of the water samples tested by the class contained the most D.O.? The least? Based on what your instructor has told you about the sources of the samples and ecological differences between them, do these differences in D.O. make sense? Which of the aquatic ecosystems sampled could sustain the healthiest community of organisms, based on D.O. alone?

2. What is the advantage of using percent saturation rather than a simple measure of D.O.? What effect does temperature have on D.O.?

THE B.O.D. TEST

As discussed above, B.O.D., or biochemical oxygen demand, is a measure of the amount of organic material present in a water sample. Because one of the goals of effective wastewater (sewage) treatment is the reduction of organic material, wastewater treatment plants usually measure B.O.D. at several stages of the treatment process.

One type of instrument used to easily measure B.O.D. of a water sample is the Hach® B.O.D. Respirometer (Fig. 9.2). This apparatus has been set up prior to your laboratory period, and your instructor will demonstrate it to you. In the Hach® B.O.D. procedure, premeasured amount of sewage or wastewater is placed in a closed bottle and connected to a **manometer,** a device for measuring oxygen pressure. Over several days, decomposer bacteria in the sample use the oxygen in the sample to oxidize organic matter, resulting in a drop in D.O. This change in gas pressure registers on the manometer scale in mg/liter B.O.D.s of two different

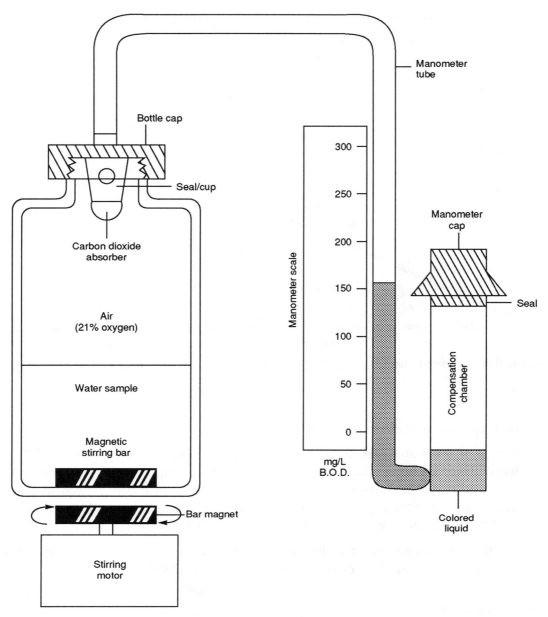

Figure 9-2. The Hach® B.O.D. Respirometer.

samples, and the lab instructor will show you how to read the scale on the respirometer. Raw sewage typically has a B.O.D. of 100–200 mg/liter. Because B.O.D. changes during the test period (usually five days), graphing the B.O.D. of a sample ordinarily results in a curve like the one in Fig. 9.3.

Your instructor may elect to perform B.O.D.s without the use of a respirometer. B.O.D. measurements may be accomplished simply by collecting a water sample in each of two 300 mL B.O.D. bottles. Be very careful to avoid the presence of air bubbles in water destined for oxygen determinations. Air bubbles derived from atmospheric air can drastically alter oxygen readings. After collecting the two samples, you should then determine the D.O. of one of the samples. Incubate the second sample in a dark place at approximately 20°C. At the end of five days (standard

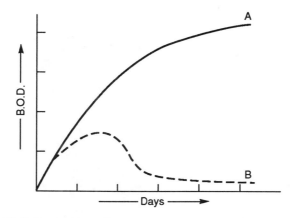

Figure 9–3. Plots of B.O.D. over time. If curve A is an example of a typical plot of a B.O.D. test run at 20°C, what would curve B probably indicate?

test time), measure the D.O. and subtract it from the D.O. reading of the first sample. This will be your B.O.D. measurement in ppm. For example, if the D.O. was 6 in your first sample and was 3 in your 5-day-old sample, the B.O.D. would be 3.

In many cases, the organic matter present in a sample of water will be excessive and you will need to dilute both samples with plain water. Trial and error will tell you the proper dilution factor to use. Common dilutions are 5%, 10%, 25%, and 50%. If you don't dilute the samples, your second sample of water may become anaerobic before the test period of 5 days is completed. Remember that a "dirty" sample of water could easily have a B.O.D. of 90–100 and a D.O. of 2. Without dilution, the sample would quickly become anaerobic. Dilution allows you to complete the 5-day test. NOTE: When calculating the B.O.D., you must take into account the dilution factor. You can do this by using the following equation:

$$BOD\ (ppm) = \frac{DO1 - DO2}{DF}\ where$$

DO1 = dissolved oxygen in sample measured minutes after dilution

DO2 = dissolved oxygen in sample measured after incubation

DF = decimal fraction of sample used

Example: $\frac{3-1}{10}$ (10% sample)* $= \frac{2}{.10}$ $=$ 20 ppm

---

*A 10% sample would contain 100 mL of actual sample diluted with 900 mL of pure water.

10

Water Quality Testing III: pH

EQUIPMENT NEEDED

1. Wide range pH (3–10) test kit, colorimetric or pH (authors use La Motte wide range pH Comparator—La Motte Chemical)

Scientists measure and compare the acidity of different solutions using the **pH scale** (Fig. 10-1), a measure of the relative concentrations of hydrogen ions (H^+) and hydroxide ions (OH^-) in the solution. A solution that is neutral (such as pure water) has equal amounts of hydrogen and hydroxide ions and has a pH of 7. Solutions with more hydrogen than hydroxide ions are **acidic** and have a pH between 7 and 0. The lower the pH, the more acidic the solution. Some examples of acidic solutions are orange juice and vinegar. Solutions with greater concentrations of hydroxide than hydrogen ions are **alkaline**, or **basic**, and have a pH between 7 and 14. The higher the pH, the more alkaline the solution. Alkaline solutions include detergents and ammonia.

Each whole-number increase or decrease on the pH scale represents a tenfold increase or decrease in acidity or alkalinity. For example, lemon juice with a pH of

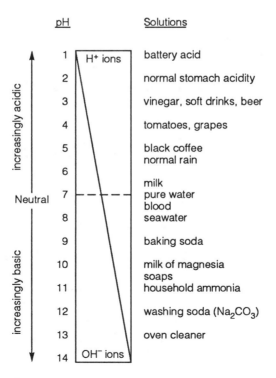

Figure 10–1. The pH scale.

approximately 2 is 10 times more acidic than wine (pH = 3) and 100 times more acidic than tomatoes or grapes (pH = 4).

Even unpolluted rainwater is slightly acid (pH about 5.6) because it reacts with natural carbon dioxide (CO_2) in the atmosphere to form carbonic acid. Most natural waters range from pH 4 to pH 9, and fish breed best at pH ranges between 6.5 and 8.5. At pH levels below 4, most animals, plants, and microorganisms perish.

One of the most serious and widely publicized pollution problems in many parts of the world, including North America, is acid deposition, often referred to as "acid rain." When the by-products of fuel combustion such as sulfur dioxide (SO_2) and nitrogen oxides (NO_x) enter the atmosphere, they form tiny droplets and particles of sulfuric and nitric acid. Acid deposition can have a pH ranging from about 1.9 to 5.5; however, the pH of most "acid rain" falls between 4.0 and 5.0. When this acid is deposited on the surface of the earth, it can result in serious environmental consequences. Acid precipitation can destroy aquatic life, kill beneficial soil bacteria, damage forests and crop plants, and leach plant nutrients from the soil. In addition, acid can react with other pollutants, such as aluminum and mercury, to produce extremely toxic new compounds. Some lakes and streams in the United States, including those in the Adirondack Mountains in the east, are so seriously polluted by acid deposition that they are virtually devoid of most forms of aquatic life.

LABORATORY EXERCISE

One measures the pH of several known and unknown solutions using a wide range (3–10) pH kit (commonly **colorimetric**) or pH paper.

1. Fill test vial to line with water.

2. Add ten drops of pH reagents.

3. Shake, then compare the color of the sample to the colors on the kit to determine the pH.

4. Fill in the chart:

Substance	pH

EQUIPMENT NEEDED

1. In the event of inclement weather, a 35mm slide series on a community sewage plant—primary, secondary, and possible tertiary treatment. If possible, use your own community sewage plant.

Wastewater or sewage is basically the flow of used water from a community. Wastewater is typically over 99% water by weight. The remainder is solid material either dissolved in the water or suspended as particulates. Sewer systems receive both domestic sewage and stormwater runoff. Raw sewage typically has a B.O.D. of 100–200 mg/L. It also contains significant amounts of phosphorus and nitrogen.

In general, the goal of wastewater treatment is to transform raw sewage into water clean enough to be discharged into a lake or stream without deleterious environmental or ecological consequences. Effective sewage treatment combines physical, chemical, and biological processes to accomplish several goals: (1) to reduce "aesthetic pollution" (unsightly debris and smelly organic material); (2) to kill pathogenic microorganisms and remove toxic wastes, both of which can cause disease; (3) to reduce organic material, or B.O.D.; and (4) to remove inorganic nutrients (nitrogen and phosphorus) which can lead to artificial eutrophication. It is important not to confuse wastewater or sewage treatment with water purification. Sewage treatment plants are not designed to purify the drinking water supply, but to reduce the potential of sewage to pollute aquatic ecosystems.

A typical wastewater treatment plant subjects sewage to a two-step cleaning process. Primary treatment involves the separation of the solids from the liquids in the raw sewage. It is a mechanical process in which screens, scrapers, and settling tanks are used to filter out debris and settle out suspended organic solids, called sludge. On average, primary treatment reduces the B.O.D. of sewage by 30%.

The second phase, called secondary treatment, is primarily a biological process. Secondary treatment is designed to further reduce the B.O.D. of the organic material. It makes use of devices that agitate and aerate the sewage in order to promote the activity of naturally occurring aerobic, decomposer bacteria. This stage results in the sedimentation of more sludge. Secondary treatment typically reduces B.O.D. by an additional 60% over primary treatment, for a net reduction in organic material of about 90%. The sludge produced in primary and secondary treatment may be used as an inexpensive source of agricultural fertilizer.

After secondary treatment, wastewater still may contain high percentages of phosphorus, nitrogen, and other materials resistant to biological breakdown. The removal of any of these substances requires highly complex and specialized chemical and physical procedures, all of which fall under the heading of tertiary or advanced treatment. Tertiary treatment can take on many forms, including precipitation, adsorption, and reverse osmosis, depending on the needs of that area. Tertiary processes add significantly to the cost of municipal sewage treatment; consequently, this is not a commonly used form of sewage treatment.

In addition to the phases described above, sewage is typically chlorinated to kill various pathogens, including bacteria and viruses.

A "TYPICAL" WASTEWATER TREATMENT PLANT IN A U.S. COMMUNITY

Primary treatment uses an aerated grit removal tank and round settling tanks (Fig. 11-1). Grit removal separates the inorganic solids that cannot be treated at this plant. The primary settling tanks use gravity to allow organic solids to settle out of suspension. The sludge is then sent to another location for screening and aerobic digestion.

Secondary treatment is accomplished through the use of bacterial decomposition aerobic tanks. The biological activity of decomposers is promoted by injecting large quantities of air that mixes and aerates the sewage. Decomposition of the solids occurs in the aerobic digestion tanks. Breakdown of the organic material in the liquid is facilitated by the addition of activated sludge (sludge rich in bacteria) and occurs in the aeration tanks. The organisms are settled out of the liquid in the settling tanks. Finally, chlorine is frequently added to the plant effluent for pathogen elimination.

Sludge is shunted to the aerobic sludge digestion tanks at several stages of the treatment process. After digestion, this rich organic material is used to fertilize selected agricultural fields in the area. The material is sprayed onto the fields used to produce agricultural crops.

If time and weather permit, your instructor may arrange or conduct a guided tour of your community treatment plant. In the event that this cannot be arranged, you may see a detailed slide presentation of the plant and various stages of the treatment process.

ACTIVITY

Construct a "flow chart" of your sewage plant and discuss functions of each component. See Figure 11-1 for an example of a typical "flow chart."

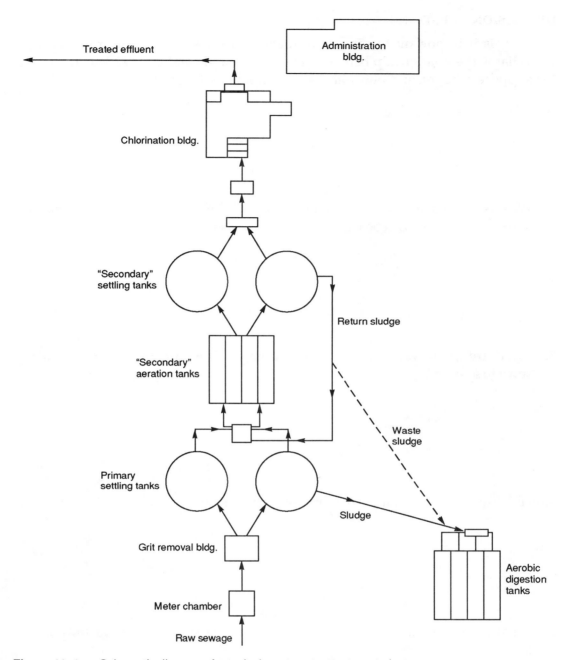

Figure 11–1. Schematic diagram of a typical wastewater treatment plant.

DISCUSSION QUESTIONS

1. Why is it important to remove nitrogen and phosphorus from raw sewage? What is the relationship between these compounds and B.O.D.? Where do you suppose nitrogen and phosphorus come from in your area?

2. Why is the wastewater treatment effluent disinfected? Specifically, what are some diseases that can be transmitted via poor sanitation?

3. What are the advantages and disadvantages of a combined storm/sanitary sewage system?

4. How might your community improve on the treatment process?

5. What are the advantages/disadvantages of using sludge from wastewater for fertilizer?

12 The Human Respiratory System

EQUIPMENT NEEDED

1. Prepared microscope slides
2. Vitalometer (respirometer)
3. 35mm slide series on lung histology and diseases are useful

In order to fully understand the impact that air pollution can have on human health, it is important to know something about the structure and function of the healthy respiratory system. In this laboratory, you will (1) familiarize yourself with the gross anatomy of the respiratory tract; (2) examine histological (microscopic) preparations of respiratory tissue; and (3) learn to measure **vital capacity,** one index of respiratory health.

RESPIRATION

Respiration itself can be defined in two different ways. At the cellular level, respiration is a complex series of metabolic reactions involving the oxidation of organic molecules, such as carbohydrates, to release energy and fuel body processes. Cellular respiration in multicellular organisms, including humans, requires the expenditure of oxygen (O_2) and results in the release of carbon dioxide (CO_2).

The second definition of respiration is the physical process of gaseous exchange between the cells of an organism and its external environment. This includes **ventilation,** the process of getting air in and out of the lungs, as well as **diffusion,** the actual exchange of oxygen and carbon dioxide between the lungs and the circulatory system and from the blood to the various tissues of the body. It is this second definition that we will mainly be concerned with in this lab.

GROSS ANATOMY OF THE HUMAN RESPIRATORY TRACT

Because only a fraction of their cells are in direct contact with the external environment, numerous organisms, including humans, have evolved specialized organs of gaseous exchange called lungs. In addition to the lungs, the human respiratory system consists of a series of passageways leading to the lungs and chest structures for moving air in and out.

The mouth and nose communicate with the lungs through a series of special structures (see Fig. 12-1). The **glottis** is an opening in the floor of the **pharynx,** protected above by a lid or **epiglottis** and supported by a cartilaginous framework, the **larynx** or voice box. The latter connects to a flexible tube, the **trachea** or windpipe, that extends into the thorax and forks into two **bronchi,** one to each lung. The trachea and bronchi are reinforced against collapse by rings of **cartilage.** In the lungs

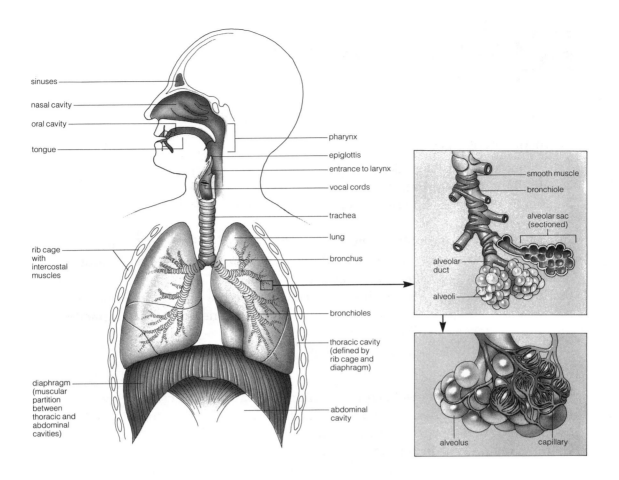

sinuses
nasal cavity
oral cavity
tongue
pharynx
epiglottis
entrance to larynx
vocal cords
trachea
lung
rib cage with intercostal muscles
bronchus
bronchioles
thoracic cavity (defined by rib cage and diaphragm)
diaphragm (muscular partition between thoracic and abdominal cavities)
abdominal cavity

smooth muscle
bronchiole
alveolar sac (sectioned)
alveolar duct
alveoli
alveolus
capillary

Figure 12–1. Human respiratory system. (After Cecie Starr, *Biology: Concepts and Applications.* Belmont, CA: Wadsworth, 1991, p. 410.)

the bronchi branch into many smaller **bronchial tubes** and **bronchioles,** each with successively thinner walls. Each bronchiole lacks cartilaginous plates and ends in a saclike atrium, having on its surface many small irregular chambers, the **alveoli** or air sacs. The alveoli are surrounded by blood capillaries, where the gaseous exchange occurs. The human lungs have 500,000,000 or more alveoli.

The cells lining the passageways of the respiratory system are equipped with microscopic hairlike organelles called **cilia.** These structures, along with **mucus** secreted by specialized **mucus glands** also lining the respiratory tract, assist in keeping the passageways free from dust and other foreign material. The cilia bend together in a coordinated fashion, moving mucus and debris up and out of the lungs. Smoking and other forms of air pollution can damage the cilia, jeopardizing the ability of the respiratory tract to cleanse itself.

In addition to the respiratory structures mentioned above, certain specialized muscles are also essential to the breathing process. The **intercostal muscles,** located between the ribs, allow the ribcage to expand and contract. The most important respiratory muscle, however, is the **diaphragm,** a dome-shaped muscle just under the lungs and above the stomach and other abdominal muscles. When the diaphragm contracts, it creates negative pressure within the ribcage, pulling air into the lungs (**inhalation**). When the diaphragm is in the relaxed position, air is expelled (**exhalation**).

Your instructor will point out the gross anatomical features of the human respiratory system using diagrams and models. Be sure to familiarize yourself with these structures before going on to the next phase of the lab.

LUNG HISTOLOGY

1. Obtain prepared microscopic slides of cross sections of mammalian lung tissue.

2. Using Figure 12-2, locate the following structures: bronchial tubes, bronchioles, alveoli, cartilage, veins, and arteries.

3. Label these structures in Figure 12-2. Be sure you understand the form and function of the identified structures.

VITAL CAPACITY AND RESPIRATORY VOLUMES

Respiratory physiologists measure volumes of air moved during each phase of the human respiratory process as indices of healthy respiratory functioning. The first of these, **tidal volume,** is the amount of air inhaled or exhaled during normal breathing. Tidal volume depends on many factors, including body size, age, health, and activity of the subject at the time it is measured. On average, human tidal volume is about 500 mL. **Vital capacity,** on the other hand, is the amount of air that can be forcibly, consciously exhaled after a maximal inhalation: 3,100 mL for women and 4,600 mL for men, on average. Body size will affect your personal reading. Vital capacity can be increased by vigorous exercise and decreased by air pollution or respiratory disease.

Even after a person expels as much air from the lungs that he or she can forcibly exhale, there is still a small volume (approximately 1 L) remaining. This leftover air is called **residual volume.**

Your instructor will demonstrate proper use of an instrument for measuring vital capacity, called the **vitalometer.** Each student will be asked to measure his or her vital capacity (taking care to use sterile mouthpieces for each test).

1. Do the vital capacities of men and women appear to differ, on average, in your class? Would you expect vital capacity to be related to height, weight, or other physical characteristics?

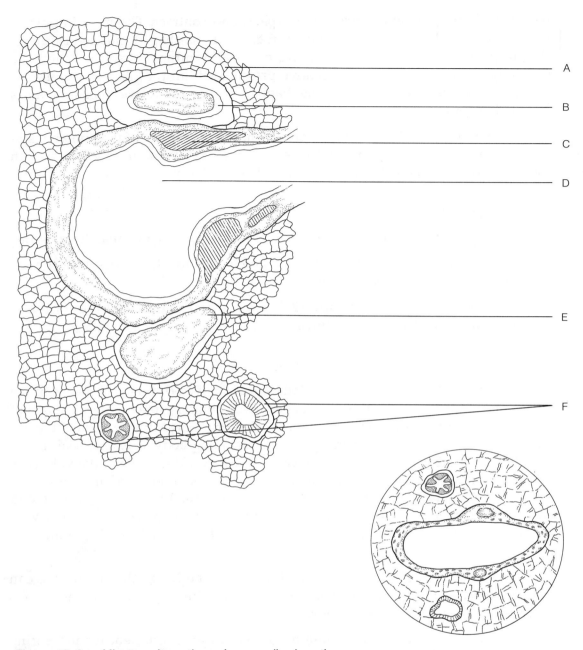

A

B

C

D

E

F

Figure 12–2. Microscopic sections of mammalian lung tissue.

2. What is the range (smallest to largest) of vital capacities in your class? How do you account for the variability you see?

3. Do the vital capacities of smokers/nonsmokers, athletes/nonathletes appear to differ substantially in your class? Why?

RESPIRATORY AILMENTS

Most respiratory diseases can be aggravated, if not directly caused, by air pollution, including cigarette smoke. **Bronchitis,** the inflammation of the bronchial tubes, can be caused or made worse by pollutants such as sulfur dioxide, which can damage cilia and cause a buildup of mucus, ultimately leading to serious infection. **Emphysema,** a disease characterized by deterioration of the bronchioles and alveoli, may be caused by air pollution. It is incurable, for the most part, and can lead to progressive oxygen deprivation and, eventually, death.

Lung cancer, one of the increasingly prevalent diseases in the United States, may be caused by carcinogenic air pollutants, many of which can be found in cigarette smoke. **Asthma,** an allergic reaction to airborne particles, is sometimes aggravated by pollution. In an asthma attack, the bronchial walls become inflamed and constricted, significantly decreasing vital capacity.

EQUIPMENT NEEDED

1. Pocket combs (students can supply their own)
2. Limewater
3. Drinking straws
4. Beakers for limewater

Air pollution takes the form of gases or aerosols (tiny liquid droplets or solid particles). Air pollution control may involve producing less air pollution, the best approach, or preventing pollution from getting into the air. Part I of this exercise looks at some methods of preventing pollution from getting into the outside (ambient) air.

PART I

REMOVAL OF PARTICULATES

Filtration

This method of particulate removal operates on the same principle as the vacuum cleaner. The vacuum cleaner forces dirty air through a bag that retains the dirt as the air passes through. In an industrial application, dirty smoke is passed through a cluster of bags that retain the particulates and the cleaned gas passes through. This does not mean, of course, that the gas is pollution-free.

Electrostatic Precipitator

When the air is dry, we find that static electricity is common. Combing our hair may cause the hair first to cling to the comb, then refuse to lie down. What has happened is that the comb has acquired a negative charge by removing some negatively charged electrons from your hair, which then has a positive charge. Because opposite charges attract, the hair is attracted to the comb; but because like charges repel, all of the positively charged hairs repel each other. Static cling likewise is due to the attraction of oppositely charged materials. Even materials that have not been charged may be attracted to a charged object.

An electrostatic precipitator utilizes this principle. When smoke passes electrically charged plates, the particles in the smoke are attracted to the plates, from which they can be collected. You can test this with a comb and some bits of paper.

Run a comb briskly through your hair several times, and hold it above some bits of paper on the table. What does the paper do?

You can also try this with cigarette ash collected from an ash receptacle.

REMOVAL OF GASES

The most common way that gas of any kind is removed from smoke is by a wet scrubber, which removes sulfur dioxide (SO_2). The sulfur dioxide reacts with crushed limestone ($CaCO_3$) according to the following equation:

$$O_2 + 2SO_2 + 2CaCO_3 \longrightarrow 2CaSO_4 + 2CO_2$$

As a result of this reaction, the sulfur dioxide oxidizes and then combines with calcium to form calcium sulfate, which is the same chemical as gypsum, widely used as drywall and plaster of Paris.

We can simulate this reaction with two different materials, the carbon dioxide in our breath and limewater [$Ca(OH)_2$]. The reaction is different, but the principle is the same. The reaction is as follows:

$$CO_2 + Ca(OH)_2 \longrightarrow CaCO_3 + H_2O$$

In this case, calcium carbonate is produced by the reaction, instead of calcium sulfate.

Blow slowly through a straw into some limewater.

What do you see?

Where did the carbon dioxide go?

THE GREATER DETROIT RESOURCE RECOVERY AUTHORITY

The city of Detroit operates the Greater Detroit Resource Recovery Authority, referred to locally as the "Detroit Incinerator." Those opposed to the construction and use of this facility were mainly concerned about air pollution. The emissions from the incinerator were measured in April 1990.

Complete the following table to evaluate the performance of the air pollution controls at the GDRRA.

Type of Pollution	Permitted	Units Emitted	Emitted as % of Permitted
Particulates	1934.40	171.36	_____
Hydrogen chloride	14,112.00	7,665.60	_____
Fluoride	180.00	10.10	_____
Dioxins and furans	0.21	0.08	_____
Mercury	3.36	3.84	_____
Cadmium	4.08	0.77	_____
Chromium	0.77	0.36	_____
Lead	65.76	4.18	_____
Carbon monoxide	7,152.00	830.00	_____
Sulfur dioxide	21,940.00	2,116.80	_____
Nitrogen oxides	10,848.00	6,240.00	_____

How closely did the incinerator's emissions compare with those permitted?

The incinerator exceeds the permit in mercury emissions. What would we need to know to decide whether that was a more serious problem than if, say, carbon monoxide exceeded the emissions permitted?

PART II

Smoking is a major contributor to indoor air pollution. After all, it involves leaf-burning, something that has been banned outdoors in many communities for many years.

On page 69 is a survey on smoking. Make 20 copies of it, and ask friends, family members, and other acquaintances to fill them out. Compile the data you collect, using the following table.

Comparing smokers and nonsmokers

No. who smoke now _____ % of total _____

No. who do not smoke now _____ % of total _____

Total _____

Comparing smokers and quitters

No. who smoke now _____ % of total _____

No. who have quit smoking _____ % of total _____

Total _____

Desiring to quit

No. of smokers wanting to quit _____ % of total _____

No. of smokers not wanting to quit _____ % of total _____

No. of smokers with no opinion on quitting _____ % of total _____

Total _____

Started smoking

Age started smoking _____

Use the grid in Figure 13-1 to make a graph to show how many people started smoking at each age. (Graphing the data for the other questions might be interesting, too.) At what age(s) did most people start smoking?

What is the oldest anyone started?

Do your data suggest why?

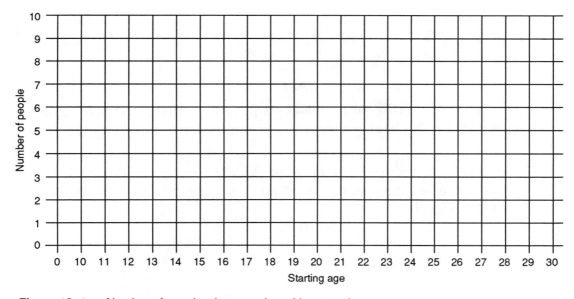

Figure 13–1. Number of people who started smoking at various ages.

TOBACCO USE SURVEY

This is a survey for the purpose of gathering information on tobacco-use habits of people in
_____School or in the community of_____.

We would appreciate your honest candid answers to the questions. We may share the results with interested people, but individual responses will be kept anonymous.

Do you now smoke cigarettes?　　Cigars?　　A pipe?　　Use snuff?

Did you ever smoke cigarettes?　　Cigars?　　A pipe?　　Use snuff?

If the answer to any of the above is yes, please answer the following questions.

At what age did you start smoking?　　Using snuff?

Why did you start?

Would you like to quit?　　If you have quit, how long did you use tobacco?

How long ago did you quit?　　Why did you quit?

What is your opinion of your school's or your community's smoking policy?

HOW ADDICTED TO NICOTINE ARE YOU?

	0 points	1 point	2 points	Score
1. How soon after you wake up do (did) you smoke your first cigarette?	After 30 min.	Within 30 min.		_____
2. Do (did) you find it difficult to refrain from smoking in places where it is forbidden, such as the library, theater, doctor's office?	No	Yes		_____
3. Which of all the cigarettes you smoke(d) in a day is (was) the most satisfying?	Any other than the first one in the morning	The first one in the morning		_____
4. How many cigarettes a day do (did) you smoke?	1–15	16–25	More than 25	_____
5. Do (did) you smoke more during the morning than during the rest of the day?	No	Yes		_____
6. Do (did) you smoke when you are (were) so ill that you are (were) in bed most of the day?	No	Yes		_____
7. Does (did) the brand you smoke have a low, medium, or high nicotine content?	Low (0.4 mg)	Medium (0.5–0.9 mg)	High (1.0 mg)	_____
8. How often do (did) you inhale the smoke from your cigarette?	Never	Sometimes	Always	_____

TOTAL POINTS _____

If you scored 4 points or more and want to quit smoking, you may need medical help to quit and should see your doctor. The 8-question questionnaire in the last part of this survey was provided by Marion Merrell Dow Inc., Kansas City, MO.

Thank you for your cooperation.

14   Thermal Inversion

EQUIPMENT NEEDED

1. Inversion chamber
2. Ice
3. Smoke generator or cigarettes

In certain parts of the country, particularly urban areas or regions characterized by hills and valleys or coastal mountain ranges, the detrimental effects of air pollution can be aggravated by an atmospheric condition called a **thermal inversion** or temperature inversion.

Under normal atmospheric conditions, air close to the earth's surface becomes heated by the sun. Warm air molecules are moving faster than those in cold air, thus warm air is lighter and less dense. The warm layer next to the earth rises gradually into the cooler layers in the upper atmosphere, carrying with it smog generated at the surface (see Fig. 14-1a). In other words, pollution is dissipated by the circulation of air during normal atmospheric temperature patterns.

When situated in a valley or near a coastal mountain range, a city is particularly susceptible to cool air from the surrounding regions creeping into the area, particularly at night when winds often shift and the earth's surface gradually cools. Because this cooler air is heavier than the warm air at the surface, the cool air displaces the warm layer upwards, inverting the normal pattern of atmospheric layers. The inversion layer prevents circulation, forming a "thermal lid" on any pollution generated at the surface (see Fig. 14-1b). This situation persists until some force reverses the temperature inversion. Where large quantities of smog are being produced by factories or automobiles, a thermal inversion can concentrate air pollutants to dangerous levels.

Major cities that are notably susceptible to thermal inversions include coastal cities, like Los Angeles, and valley cities. Some of the worst air pollution disasters

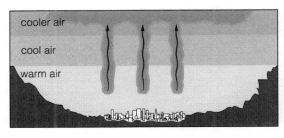

a Normal pattern

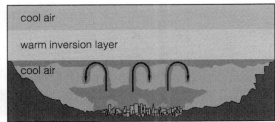

b Thermal inversion

Figure 14–1. Thermal inversion traps pollutants in a layer of cool air that cannot rise to carry the pollutants away. (From G. Tyler Miller, Jr., *Living in the Environment*, 7th ed. Belmont, CA: Wadsworth, 1992, p. 584.)

have been a result of persistent thermal inversions. One such incident was the so-called "killer smog" in 1948 in Donora, Pennsylvania, which resulted in a number of pollution-related deaths and illnesses.

LABORATORY EXERCISE

Your instructor will demonstrate a simulated thermal inversion for the class.

[Some biological supply houses sell complete inversion chambers, and you may wish to purchase one; however, a temperature inversion chamber is easy to construct. Simply take a cylinder of clear plastic (about 45 cm high and open at both ends), and drill a hole in the cylinder about 5 cm from the bottom. Insert a piece of rubber tubing. Hang two thermometers inside the cylinder (one on the top and one toward the bottom). Place the cylinder in a container of ice (5–6 cm deep). The ice should create two different temperatures (an inversion) in the cylinder. Using rubber tubing, blow smoke (smoke generator or cigarette) into the cylinder. Smoke should rise only to inversion level. When this is clearly visible, remove ice and put in hot water, which will "break" the inversion, and smoke will no longer be contained and will flow out the top of the cylinder.]

15   Introduction to the Arthropods

EQUIPMENT NEEDED

1. Preserved specimens of *Procambarus*, any centipede or millipede species, *Romalea microptera*, *Argiope*, *Dermacentor variabilis* or other tick species
2. Different insects showing incomplete and complete metamorphosis. Plastic biomounts are useful.

The majority of known animal species are in the phylum Arthropoda. Over a million species of insects alone have been identified. The biomass of insects outweighs that of human beings. All arthropods have a segmented body, sclerotized exoskeleton, and paired jointed appendages. The versatile exoskeleton provides protection, prevents desiccation of terrestrial arthropods, and aids in mobility by serving as an attachment point for muscles. A problem with the exoskeleton is that it can't grow with the animal; consequently, it must be shed (molted) occasionally.

Because arthropods constitute a significant component of soil fauna, you will likely encounter numerous types of arthropods when examining soil samples. To aid in your understanding and appreciation of the various arthropods that will be seen in your soil samples, selected preserved specimens are available for study.

CLASS CRUSTACEA

Crustaceans are primarily marine organisms and include representatives such as crabs, shrimp, lobsters (Figure 15-1), and copepods. However, some crustaceans are terrestrial. In some of your soil samples, you will probably see some terrestrial isopods (wood lice or sow bugs). Although it is not terrestrial, you should examine a crayfish (*Procambarus*) to observe arthropod and crustacean characteristics. Review the following arthropod characteristics that a crayfish exhibits:

1. **Exoskeleton**

2. **Segmentation**

3. **Grouping of segments**—Note two major body regions, which are the cephalothorax (head merged with thorax) and abdomen. The cephalothorax may appear to be a solid region with few traces of segmentation. Actually, the thorax is covered by the **carapace,** which is simply a backward extension of the head.

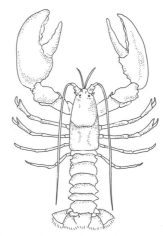

Figure 15–1.   Example of Crustacea.

4. **Paired jointed appendages**—Note the largest leg or **cheliped** (pinchers). How many pairs of legs do you see? Observe other paired jointed appendages such as mandibles (jaws) and first and second antennae.

Use the pictorial key in Figure 16-2 to recognize the key characteristics of the various arthropod classes.

CLASS CHILOPODA

The Chilopoda or centipedes (Figure 15-2) are terrestrial and breathe by means of air tubes. As with the crayfish, note the sclerotized cuticle, segmented body, and paired jointed appendages. The body is wormlike with head and trunk. How many legs per segment do you see? What about antennae? A pair of poison claws, located on the first segment behind the head, is

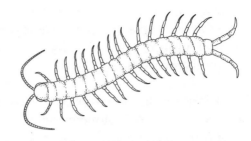

Figure 15–2. Example of Chilopoda.

used to paralyze insects and other small animal life, which constitutes the principal source of food. The class as a whole is considered beneficial; however, some of the larger forms are capable of inflicting bites painful to humans. In temperate climates, centipedes usually are not more than 4 cm in length, but they may reach a meter or more in the tropics. The house centipede is often seen in American homes. Its food commonly consists of other insects, and it is considered harmless to humans.

CLASS DIPLOPODA

The Diplopoda are known as millipedes (Figure 15-3) or "thousand legs." They don't really have as many legs as the name implies. What similarities do millipedes and centipedes possess? What are the obvious differences?

Figure 15–3. Example of Diplopoda.

How many legs per body segments do they possess? Millipedes prefer dark, damp places in soil, under logs, and in homes. When disturbed, they often curl up into a coil. Their normal food is decaying organic matter.

Question: What do centipedes and millipedes have to do with the metric system?

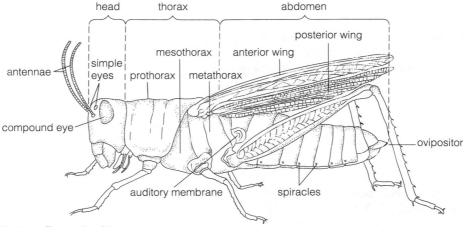

Figure 15–4. Example of Insecta.

CLASS INSECTA

The insects are a large and important class of the arthropods. Many are of economic importance, particularly with respect to agriculture and public health. You will examine a preserved specimen of *Romalea microptera* (Figure 15-4). This relatively large grasshopper is found in some of the southern states.

Review the following features of insects' external anatomy:

1. Review the main body regions. On what body region do the legs and wings originate?

2. How many pairs of antennae (Figure 15-5) do you see?

3. Compound eyes—insects do form images and have color vision.

4. Simple eyes—find the three simple eyes on the top of the head. They don't form images but do detect light and dark.

5. Mandibles—Note the jaws on the head region. Not all insects have mandibles. For example, some insects have piercing–sucking mouthparts. Do you think these insects eat the same thing as grasshoppers and other insects with "jaws"?

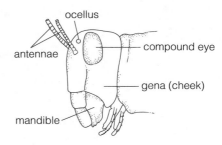

Figure 15–5. Head region of an insect.

6. Legs—Examine the individual leg segments (Figure 15-6). How does leg 3 differ from leg 1? In the grasshopper, which legs appear to be of a saltatorial (adapted for leaping) type?

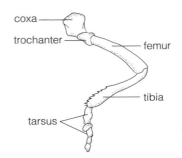

Figure 15–6. Leg segments of an insect.

7. Female grasshoppers possess an ovipositor for placing eggs in an appropriate habitat (soil, etc.). Do you have a male or a female grasshopper?

CLASS ARACHNIDA

In Arachnida, you find such organisms as spiders (Figure 15-7), scorpions, mites, and ticks (Figure 15-8). There is considerable variation in body organization. The bodies of members of this group are usually composed of a cephalothorax and an abdomen. Antennae are absent. We will examine *Argiope* (the garden spider) and *Dermacentor* (American dog tick).

In *Argiope*, observe the following:

1. Note the two major body divisions: cephalothorax and abdomen.

2. How many legs do you see? This is an easy characteristic to use for identifying adult arachnids.

3. Arachnids don't have mandibles like you saw in the grasshopper. The arachnids typically possess pincherlike or fanglike **chelicerae** and leglike **pedipalps.** With the aid of your lab instructor, identify these structures.

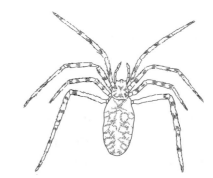

Figure 15–7. An arachnid: the garden spider, *Argiope.*

In *Dermacentor*, observe the following:

1. As with *Argiope*, note the four pairs of legs.

2. Note the **capitulum** or specialized head, which bears the chelicerae and pedipalps. A **hypostome** also occurs on the capitulum. When the tick feeds, the chelicerae and hypostome are pushed into the skin of the host. The hypostome is armed with small teeth that anchor the hypostome in the skin. This makes it difficult to remove the tick.

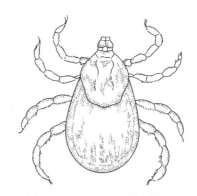

Figure 15–8. An arachnid: the American dog tick, *Dermacentor*.

METAMORPHOSIS

Most insects undergo changes in shape or form during their development. Some have minor modifications; others display drastic changes. When slight changes in appearance from the immature to the adult stage occur, the insect is said to undergo **incomplete metamorphosis** (Figure 15-9). The young resemble adults except for size, wing development, and reproductive maturity. A grasshopper exhibits this type of metamorphosis. The stages of development are egg —> nymph —> adult.

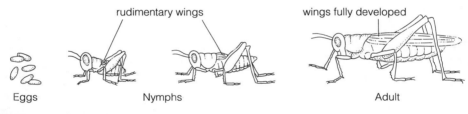

Eggs   Nymphs   Adult

rudimentary wings   wings fully developed

Figure 15–9. Incomplete metamorphosis.

Many insects display more drastic alteration in progressing from immatures to adults. Immatures are typically wormlike. The stages of development are egg —> larva —> pupa —> adult. Flies, beetles, mosquitoes, and butterflies are some examples of insects with **complete metamorphosis** (Figure 15-10).

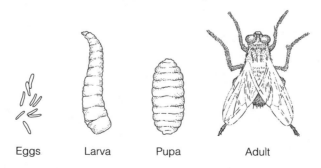

Eggs   Larva   Pupa   Adult

Figure 15–10. Complete metamorphosis.

16 Soil Diversity

EQUIPMENT NEEDED

1. Soil samples from applicable study sites
2. Berlese-Tullgren funnels
3. Buchner funnel—optional

It may come as a surprise to learn that soil contains a complex, diverse, and very important community of organisms, both large and small. The soil ecosystem plays a key role in the cycling of most nutrients throughout the biosphere. Not only does the soil community support decomposer organisms; it contains nitrogen-fixing bacteria, which convert atmospheric nitrogen to forms of nitrogen that can readily be used by plants and animals.

Soil is basically a mixture of decayed material (**abiotic,** or nonliving) and organic (**biotic,** of or produced by living organisms) substances. Rather than being a homogeneous mixture of these materials, however, the soil ecosystem is layered, or **stratified.**

Soil scientists call these layers "**horizons,**" and they often speak of the soil horizons characterizing a given geographic region as the "**soil profile**" for that locality.

In general, the **A horizon** (see Figure 16-1) contains the greatest amount of organic material. It includes **leaf litter** (dead and decaying vegetation), **humus** (plant material in an advanced state of decay), and particles of quartz and other minerals in the sand and silt size range.

The **B horizon,** on the other hand, consists mainly of colored mineral soil (inorganic material), but also it includes some organic material. Plant roots may extend down into the B horizon, and organic nutrients from the A horizon may leach down into this layer.

The **C horizon,** the deepest mineral horizon of a soil, consists of unconsolidated rock material that is transitional between the parent rock below and the A and B horizons above.

SOIL FAUNA

Organisms from all the major kingdoms are represented in the soil community, including plants, animals, fungi, and microorganisms. For the remainder of this lab we will be concerned primarily with the animal members of the community, or **soil fauna.**

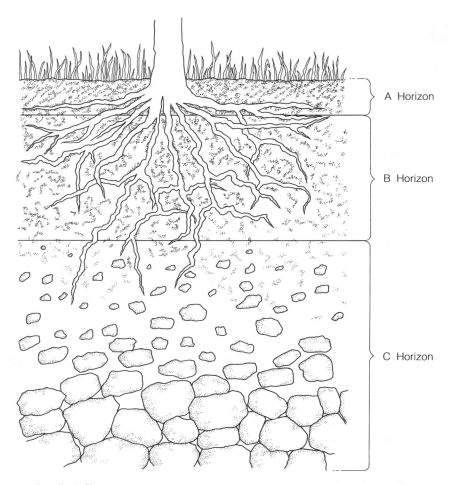

Figure 16–1. A soil profile.

Some of the most commonly encountered soil faunal organisms include **nema-todes** (roundworms), earthworms, and a variety of species belonging to the phylum Arthropoda. Soil arthropods include many-legged organisms, like isopods, centipedes, and millipedes, and eight-legged species, such as spiders and mites. The largest group within the arthropods is the class Insecta (or Hexapoda), which are six-legged animals with three body segments. These include springtails, ants, and beetles. The major arthropod groups can be distinguished by using the dichotomous key in Figure 16-2.

Soil herbivores include millipedes, mites, and springtails, which eat holes in decaying leaves. Others feed primarily on wood. Soil herbivores are among the most important producers of humus. Others, including beetles, mites, and centipedes, are consumers that prey on other soil animals.

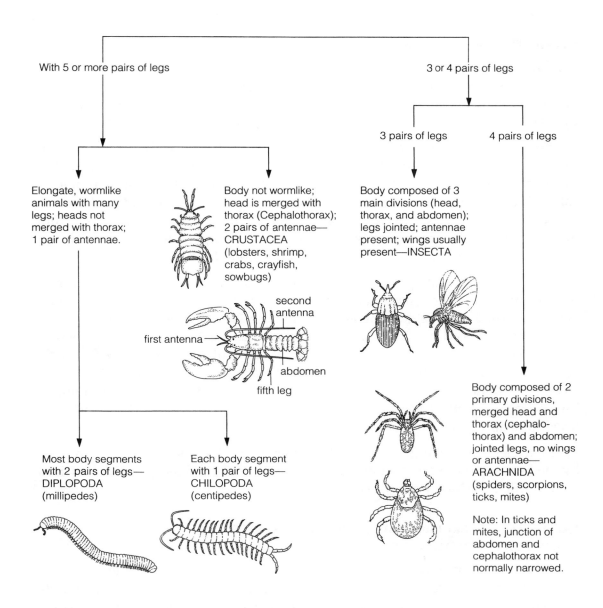

With 5 or more pairs of legs

3 or 4 pairs of legs

3 pairs of legs

4 pairs of legs

Elongate, wormlike animals with many legs; heads not merged with thorax; 1 pair of antennae.

Body not wormlike; head is merged with thorax (Cephalothorax); 2 pairs of antennae—CRUSTACEA (lobsters, shrimp, crabs, crayfish, sowbugs)

Body composed of 3 main divisions (head, thorax, and abdomen); legs jointed; antennae present; wings usually present—INSECTA

second antenna

first antenna

abdomen

fifth leg

Most body segments with 2 pairs of legs—DIPLOPODA (millipedes)

Each body segment with 1 pair of legs—CHILOPODA (centipedes)

Body composed of 2 primary divisions, merged head and thorax (cephalo-thorax) and abdomen; jointed legs, no wings or antennae—ARACHNIDA (spiders, scorpions, ticks, mites)

Note: In ticks and mites, junction of abdomen and cephalothorax not normally narrowed.

Figure 16–2. A key to the major classes of arthropods.

EXTRACTING SOIL ORGANISMS

Soil organisms, in general, live in a dark, moist, "semiaquatic" environment. Touch and smell are their most important means of orientation. Many species lack eyes but have a light-sensitive cuticle (outer covering). Many have well-developed tactile hairs and antennae.

Because the air between soil particles is usually saturated with water, most soil dwellers have a water-permeable skin or cuticle and lack the ability to prevent the evaporation of liquid from their bodies. As a result, soil organisms are physiologically restricted to their semiaquatic habitat and show behavioral adaptations for preventing moisture loss. **Negative phototaxis,** a tendency to move away from light, is exhibited by the majority of soil fauna. When given the opportunity, soil animals will move to an area with a temperature around 15°C.

Methods for collecting these species, which involve extracting them from their native soil, exploit the typical behaviors exhibited by these organisms. A common soil fauna extraction method, the **Berlese-Tullgren funnel,** uses a lamp set up over a soil sample placed in a large funnel. Mobile representatives of soil (e.g., insects, mites, spiders) travel away from the bright, warm light, seeking a more favorable environment. Eventually they drop through the open end of the funnel into a collecting jar containing water or alcohol, where they can be counted and examined. This extraction method was used to collect the samples you'll be working with, and it will probably be on display in the lab.

Many less mobile organisms, such as nematodes, are not effectively extracted by the Berlese-Tullgren funnel method. An alternate method using suction, such as the **modified Buchner funnel,** can be used for nematode extraction.

LABORATORY EXERCISE: COMPARING SOIL DIVERSITY IN DIFFERENT COMMUNITIES

Some ecological factors affecting soil faunal diversity include soil profile, nutrients, humidity, temperature, and pH. The use of synthetic pesticides to control soil pests can have a long-lasting deleterious effect on the soil community. Repeated disturbance of the soil's upper layers by plowing, as well as overfarming (repeated nutrient depletion), can also result in a significant decline in faunal diversity.

1. Obtain a prepared sample of animals that have been extracted from local forest (leaf litter), meadow, moss, and agricultural soils. (1 sample per 2 individuals)

2. Using a dissecting microscope and the arthropod key in Figure 16-2, identify as many of the different organisms present as possible.

3. Fill in this diversity chart comparing the four samples, recording the number of organisms of each class identified.

Arthropod Class	Meadow	Moss	Leaf Litter	Agricultural
Crustacea				
Chilopoda (centipede)				
Diplopoda (millipede)				
Arachnida				
Hexapoda (Insecta)				

4. Which soil type contains the greatest diversity (greatest number of different kinds) of organisms? Which is the least diverse? Which community is the healthiest, based on your findings? The least susceptible to disturbance?

5. Which types of soil organisms are most prevalent among all the samples examined? In other words, which organisms display the greatest adaptability to different soil conditions?

6. What are some environmental factors that might account for the differences in faunal diversity?

7. What are some farming practices that might be adopted to protect the soil community?

EQUIPMENT NEEDED

1. Decibel meters
2. Tuning forks: 128; 1,024; 4,096 cps (approximate)
3. Audio-frequency generator—optional
4. Ear model

Hearing is actually a mechanoreceptive sense; the ear responds to mechanical vibrations produced by sound waves in the air. Using models and transparencies, your lab instructor will discuss components of human ear and sound transmissions pathways (see Fig. 17.1). Emphasis will be on components most susceptible to noise damage. Sound-induced hearing loss is due not solely to ill effects on the eardrum, as so many people erroneously believe. Sound-induced motion of the fluid in the cochlea induces shearing and bending movements of the hair cells in the organ of Corti, which in turn results in nerve impulses transmitted by an auditory nerve. Prolonged and excessive noise eventually produces deterioration and, finally, destruction of hair cells, and this disrupts the sound transmission mechanism.

Three essential characteristics of sound affect the perception of noise: intensity (loudness), frequency, and exposure time (period of duration).

You should be familiar with the following terminology:

a. **Decibel** (dB)—the unit of sound intensity or pressure. The softest sound that can be heard by humans is 0 dB. The average individual can distinguish about 130 decibel steps in sound intensity. One hundred and thirty dB is generally considered the pain threshold. In other words, 130 dB is normally considered the greatest pressure that the ear can interpret as sound instead of pain. Extended exposure (8 hours) to 100 dB can result in hearing damage. A power lawn mower produces about 100 dB. Levels such as 150 dB can rupture your eardrums. Busy urban traffic emits about 70–80 dB; normal conversation is about 50–60; a whisper is about 20–30. Sound levels of 125 dB have been noted with highly amplified rock music. Your lab instructor may provide you with a table containing additional comparisons.

b. **Sound frequency** All noises send out sound waves that vibrate at various frequencies. The number of vibrations per second—the number of times per second that the sound waves emitted exert a pulsating pressure (sound pressure) on the ear—is the frequency of the sound, usually described in cycles per second (cps) or Hertz (Hz). A healthy human ear normally responds to frequencies ranging from a low frequency level of 20 cps to a high frequency level of 20,000 cps. Ultrasound is above 20,000 cps (silent dog whistles, etc.).

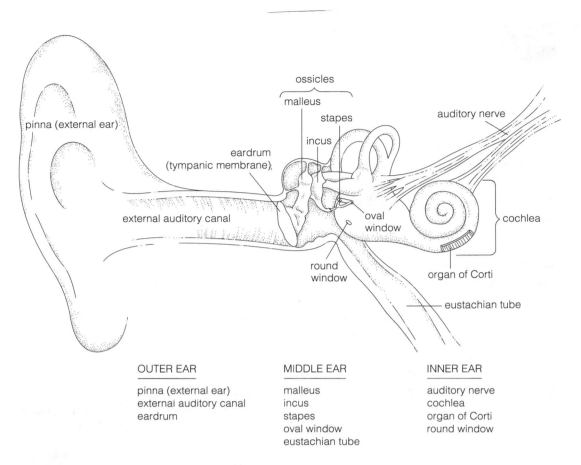

OUTER EAR	MIDDLE EAR	INNER EAR
pinna (external ear)	malleus	auditory nerve
external auditory canal	incus	cochlea
eardrum	stapes	organ of Corti
	oval window	round window
	eustachian tube	

Figure 17–1. Structures of the ear and the pathway of sound transmission. Sound waves reach the outer or *external ear* and vibrations travel through the external *auditory canal* to the eardrum or *tympanic membrane*. Vibrations are transmitted to the three *ossicles* (bones) of the middle ear—the *malleus* (hammer), the *incus* (anvil), and the *stapes* (stirrup). The stapes rests against the *oval window*. It transmits vibrations to the *cochlea*, which is filled with a liquid called *perilymph*. Vibrations travel through the ducts of the cochlea and bend small *hair cells* connected to a membrane (the membrane and the hair cells make up the *organ of Corti*). The mechanical impulse is now changed to a nerve impulse and is transmitted to the brain by way of the *auditory nerve*. The *round window* serves to prevent the feedback of vibrations to the middle ear by bending the opposite way from the oval window.

Intensity of sound is not the only factor involved in noise-induced ear damage. The effect a certain sound intensity has on the human ear depends on the frequency of the sound. Although sensitivity is greatest between 1,000 and 4,000 Hz, it falls off for both lower and higher frequencies. Because at high intensity levels certain sound frequencies are more damaging than others, it might be important to know the frequency distribution of the noise to which a person might be exposed.

Because we tend to be more sensitive to high-pitched sounds (above 1,000 cps) than to low-pitched sounds, the most common sound pressure scale is weighted for high-pitched sounds. It is called the A scale; consequently, dB units are normally expressed as "dBA."

EXERCISES

1. Working in groups, measure various sound levels with a decibel meter. This type of instrument is commonly used in industry and government. Hearing loss is currently costing millions of dollars a year in the form of workers' compensation and disability. Both industry and government are consequently giving increasing priority to noise pollution problems. In addition to simple dB levels, exposure time is critical. For example, there is general agreement that hearing loss begins with prolonged exposure (8 hours) to 80–90 dB at 300–9,600 Hz. However, exposure to 115 dB for just 20 minutes would have the potential for being harmful. A jet engine producing 200 dB could promote hearing loss in less than 10 minutes. There is some evidence that even 80–90 dB may be too high to protect human hearing. Some authorities recommend 70 dB as the safe limit if exposed for long periods of time on a regular basis.

2. Your lab instructor will demonstrate the use of an audio-frequency generator. Remember, the normal frequency response limit for a healthy human ear is set at 20–20,000 cps. The first sign of hearing loss is usually noted as a loss of hearing in the high frequency range.

3. Using low-, intermediate-, and high-frequency tuning forks, the following auditory exercises should be performed by students working in pairs:

   a. **Hearing Acuity**

   Working in pairs, direct your partner to close his/her eyes and to insert a finger in one ear. You should then strike the prongs of a tuning fork of low frequency (approximately 128 or 256 cps) to set the fork in motion. Hold the tuning fork by the handles and don't touch the prongs while the fork is vibrating. Place the tuning fork close to your lab partner's open ear and then move it away (slowly) in a direct line until the individual can no longer hear the sound. Note the distance. Repeat the experiment with the other ear and compare distances. To compare auditory acuity for different pitches of sound, repeat the experiment with tuning forks of intermediate frequency (1,024 cps) and high frequency (4,096 cps).

   NOTE: When performing the above experiments, be sure to use the same methodology for all tests. For example, strike the tuning fork with equal force for each test and move the fork away from each ear at the same speed.

   b. **Bone Conduction**

   Vibrations passing through the skull can set the cochlear fluid in motion and create a sensation of sound. If the handle of a tuning fork is placed against any portion of the cranium, the sound can be heard even if both ears are plugged. This is **bone conduction.** If both ears are blocked, the sound will be heard only if the tuning fork contacts the skull. Remove the tuning fork and then replace it to check that the sound is perceived only through bone conduction. Would the bone conduction test be useful in distinguishing middle-ear deafness from inner-ear deafness? Why?

EQUIPMENT NEEDED

1. Shallow boxes with window glass to cover them or large, widemouth jars such as used by restaurants (1 for each team)
2. Black cloth, black paper, or black paint
3. Thermometers (1 for each team)
4. Wood, aluminum sheet, tacks (or screws), coat hanger wire to construct solar energy concentrators—a shop class might be able to construct these

SOLAR ENERGY COLLECTOR

A familiar method of using solar energy is the collectors on roofs of homes and office buildings in some parts of the United States. Such collectors have an air space that is covered by glass. Shortwave solar radiation passes through the glass and is absorbed by the dark surfaces on the interior of the collector. The energy is given off by these surfaces in the form of longwave radiation, which cannot pass through the glass, so the temperature inside the collector rises.

In some solar systems of this type, the heated air from the collectors is circulated into the building through a bin of rocks. The rocks are heated in this way, and they provide heat storage for hours when there is little or no sunlight. At those times, air circulating past the rocks gets heat from the rocks and can heat the building. Other collectors have tubes containing water, Freon (a chlorofluorocarbon), or other materials to transmit heat into the building.

A shallow box, such as the type that a case of soda pop cans come in, can be used as a solar collector if covered with a piece of window glass. A large glass jar with a lid can also be used. Use black paper, cloth, or paint to cover the inside of the box. If you use a jar, just cover the inside of the jar from bottom to top halfway around (half of the jar's circumference).

To measure the effectiveness of your collector:

1. Take it to a sunny place outside, but leave it open (take the glass off the box or the top off the jar).
2. Read the air temperature at the collector testing site with a thermometer; then place the thermometer in the collector.
3. Be prepared to measure time and temperature. If working on a team, one person can be a timer to tell another person when to read the temperature. Check the time, and close the collector leaving the clear side toward the sun. (If using a jar, place it on its side with the unpainted side to the sun.)

4. Record the temperature at 5-minute intervals on the following chart for as long as your instructor tells you to.

5. Transfer the data to the grid in Figure 18-1 to make a graph of your results.

6. Record any observations you can make from analyzing your graph.

Time (minutes)	Temperature	Observations
0		
5		
10		
15		
20		
25		
30		
35		
40		
45		
50		
55		
60		

SOLAR ENERGY CONCENTRATOR

Like many chemical substances that are said to be "stronger" or "more powerful" when they are more concentrated, solar energy is also more powerful when it is concentrated. Children often become aware of this at an early age when someone shows them how to use a magnifying glass to focus sunlight on paper or their skin. The effect of such concentration is soon apparent!

Devices that concentrate solar energy on a larger scale include a field of large mirrors that focus the sunlight on a furnace or boiler mounted on the top of a tower, a parabolic reflector dish that focuses sunlight on a collection point (much like a satellite TV dish focuses the TV signal on an electronic device mounted in front of it), or a parabolic trough that focuses sunlight on a long tube mounted on the trough.

The last type of concentrator is easiest to duplicate for individual use. You can construct a parabolic trough concentrator that will cook a hotdog. Figure 18-2 shows what the concentrator will look like. Here's how to make one:

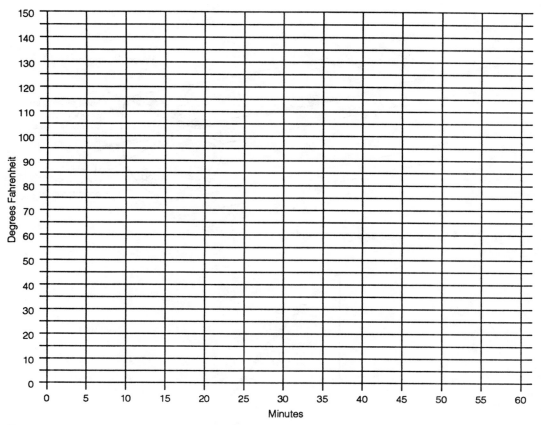

Figure 18–1. Plot of solar collector temperatures over time.

1. Cut out two pieces of wood that are 1/2 to 1 inch thick using the pattern in Figure 18-3.

2. Tack a piece of shiny aluminum to the curved edge of each piece of wood. The aluminum should be about 10 inches long and wide enough to fit around the curved part of the wood.

3. Cut a piece of stiff wire, such as coat hanger wire, an inch or so longer than your completed trough. Sand the paint off the wire.

4. Drill a hole in each wood piece exactly where the dot in Figure 18-3 is and just big enough for the wire to pass through easily.

To cook a hotdog, poke the wire through one of the wood pieces, then through the length of the hotdog, and finally through the other wood piece. Aim the trough toward the sun. Hold it by the wood, and tilt it so the sunlight is focused on the hotdog. Prop it up with a rock or other convenient object until the hotdog is done.

You might want to build a stand to support the trough. Be sure to clean off spatters when finished cooking, and keep the aluminum shiny.

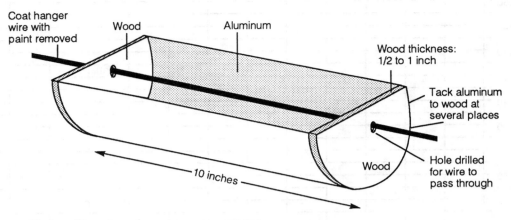

Figure 18–2. Plans for solar energy concentrator.

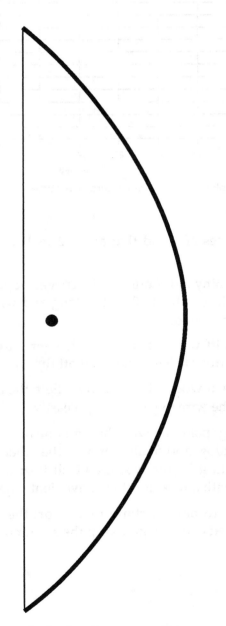

Figure 18–3. Pattern for wooden ends of solar energy concentrator.

19 **Energy Conservation**

EQUIPMENT NEEDED

None

Most of us in the United States use too much energy. Because our energy is relatively cheap, we waste natural gas, electricity, motor fuel, and other forms of energy. For example, most people keep their homes at comfortable temperatures throughout the day, when unoccupied, and at night, when their blankets would keep them warm. They readily leave lights on in unoccupied rooms and may not pay much attention to the efficiency of their major appliances and climate-control systems. Perhaps most serious of all from an environmental standpoint, they make frequent, unnecessary trips in vehicles that have very low energy efficiency.

Because most of our energy currently comes from nonrenewable energy resources, we are rapidly depleting those resources. Furthermore, use of nonrenewable energy resources has a much greater impact on the environment than use of perpetual or renewable resources would ordinarily have. Conservation of energy at the present time, therefore, would help preserve nonrenewable resources, reduce pollution, and save us money.

NATURAL GAS

Over several years, a homeowner kept track of natural gas used during the heating season, considered to be the months of October through May, when the furnace was most likely to be used. The data appear in Table 19-1. (The values are in hundreds of cubic feet, ccf, so the 1978–79 season actually used 156,700 cubic feet, but it is easier to work with 1,567 than with 156,700.)

1. Plot the data on the grid in Figure 19-1. Make either a bar chart or a line graph.

2. What major change do you observe in the 1984–85 season?

In the summer of 1984, the homeowner replaced a very old gas furnace with a new high-efficiency gas furnace, which was responsible for major energy savings. Not all of the energy savings indicated by the data resulted from the furnace, however.

At about the same time the furnace was replaced, the old thermostat was replaced with a setback thermostat. This changes the temperature, at times when

Heating Season	ccf	Heating Season	ccf
1978–79	1,567	1985–86	1,092
1970–80	1,421	1986–87	980
1980–81	1,484	1987–88	961
1981–82	1,576	1988–89	886
1982–83	1,353	1989–90	941
1983–84	1,539	1990–91	910
1984–85	1,066	1991–92	954

Table 19–1. Natural gas use for heating over 14 heating seasons (in hundreds of cubic feet = ccf).

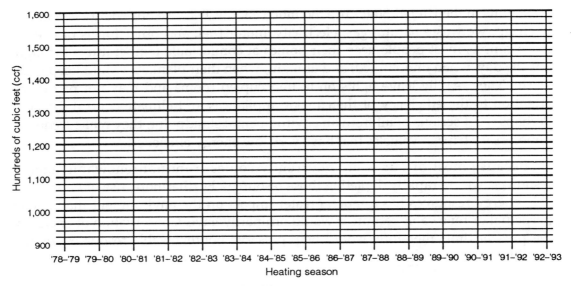

Figure 19–1. Plot of natural gas use, 1978–92.

the residents are out during the day or at night when they are in bed, to energy-saving levels that would be uncomfortable if people were up and about. Also, between June 1983 and January 1987, the windows and exterior doors were replaced with more energy-efficient ones. It is likely, though, that much of the energy and money saved was the result of the furnace replacement.

One way to determine how much effect these changes had is to average the gas consumption in the earlier years, and then to see how the gas consumption for each year compares with that average. To obtain the average gas consumption per heating season for the first 6 years, add the gas consumed and divide by 6.

Total gas consumed in heating season during 1978–1983 =_____ccf.

Average gas consumption, 1978–1983 = total gas consumed/6 =_____ccf.

Now let's see how each heating season compares with this average. To do that, complete the table that follows.

First, fill in column 3: Subtract the average gas consumption from the amount used, in hundreds of cubic feet, each season. Note that some values will be positive and others negative.

Then fill in column 4: Divide column 3 by the average gas consumption and multiply by 100. Retain plus and minus signs.

Finally, plot values from column 4 on the grid in Figure 19-2. Be sure to watch the plus and minus values.

Column 1: Heating Season	Column 2: Seasonal Gas Use (ccf)	Column 3: Seasonal Gas Use Minus Average Gas Use (ccf)	Column 4: Percent Deviation
1978–79	1,567	_____	_____
1979–80	1,421	_____	_____
1980–81	1,484	_____	_____
1981–82	1,576	_____	_____
1982–83	1,353	_____	_____
1983–84	1,539	_____	_____
1984–85	1,066	_____	_____
1985–86	1,092	_____	_____
1986–87	980	_____	_____
1987–88	961	_____	_____
1988–89	886	_____	_____
1989–90	941	_____	_____
1990–91	910	_____	_____
1991–92	954	_____	_____

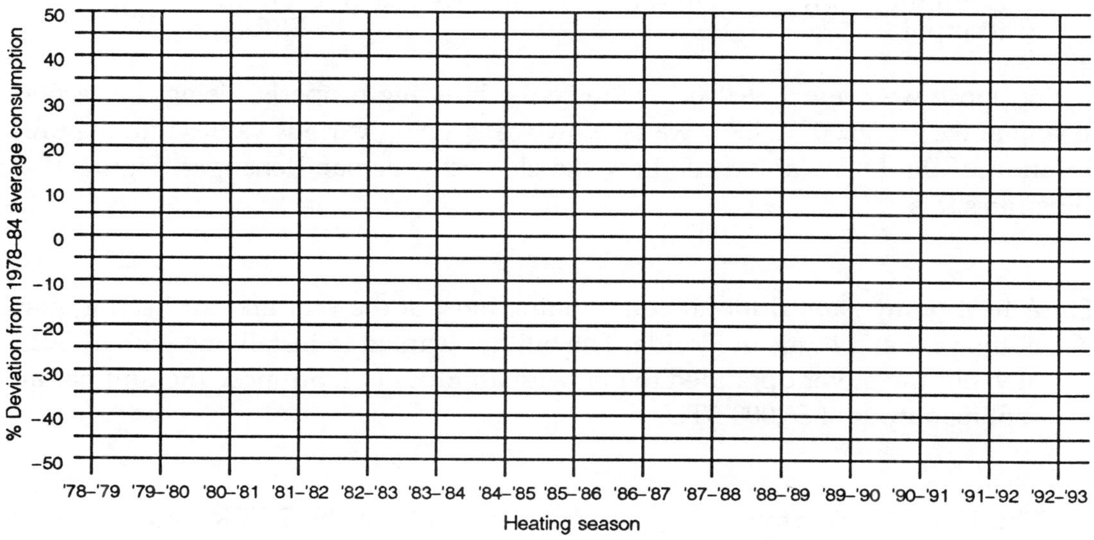

Figure 19–2. Plot of percent deviation from average natural gas use, 1978–92.

ELECTRICITY

Suppose that the EnergyGuide sticker on a new refrigerator gives an estimated annual operating cost of $53, based on a national average electric rate of 7.88¢ per kilowatt hour. According to the sticker, the model with the highest energy cost would require $88 to operate it for a year at the same price for electricity.

1. What percent more electricity does the less efficient model use? Because the amount of electricity used is related to the amount spent on electricity, the following formula would provide the answer:

   $\dfrac{\$88 - \$53}{\$53} \times 100 =$ _____ %

2. If the more efficient refrigerator costs $600, in how many years would you recover its total cost if you bought it instead of the less efficient refrigerator?

   $600/Amount saved in electricity per year = _____

   Of course, it might be better to consider how long it would take for the savings to pay for the *additional cost* of a more efficient appliance before it starts saving you money. If the price of electricity went up, the efficient appliance would save you more money.

   A heat pump provides both heating and cooling. The heating efficiency is measured by the coefficient of performance (COP) and the cooling by the energy efficiency rating (EER), which is also used with central air conditioners. The COP will not be discussed. The EER is the ratio of the BTU output (cooling) to the watt input (electricity), so it indicates directly how much electricity is needed for a certain amount of cooling.

   The formula for the EER of a unit is:

   $\dfrac{\text{BTU output}}{\text{watt input}} = \text{EER}$ To solve for watts, we use $\dfrac{\text{BTU output}}{\text{EER}} = \text{watts}$

   Suppose we needed 36,000 BTU to cool a building properly. Using the second formula, we get 36,000/EER = watts. Now we can plug various values into the EER location of the formula and see how the electricity demand changes. Figure 19-3 illustrates this.

3. A heat pump is used for air conditioning most of the year and for heating part of the year at a home in Florida. The builder wanted to install one with an EER of 9, but the buyer upgraded to one with an EER of 11. Suppose the unit has an cooling output of 36,000 BTU.

   The 11 EER unit draws _____ fewer watts than the 9 EER unit, or about _____ % as many as the 9 EER unit.

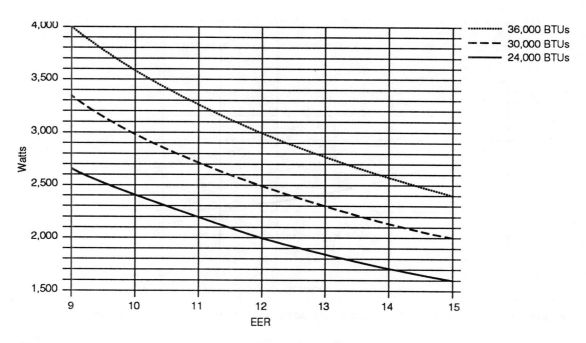

Figure 19–3. Heat pump and air conditioner efficiencies for various sized units (BTUs).

A 15 EER unit draws _____ fewer watts than the 9 EER unit, or about _____ % as many as the 9 EER unit.

GASOLINE

A major part of our energy consumption is associated with use of personal automobiles. Saving energy used in cars would not only conserve energy resources and cut down on a major source of pollution, but would save money as well.

1. Use Figure 19-4 to complete the table below. Assume you drive 15,000 miles per year.

MPG	Gallons	Price per Gallon	Cost per Year
20	_____	$1.00	_____
20	_____	$1.25	_____
20	_____	$1.50	_____
35	_____	$1.00	_____
35	_____	$1.25	_____
35	_____	$1.50	_____
50	_____	$1.00	_____
50	_____	$1.25	_____
50	_____	$1.50	_____

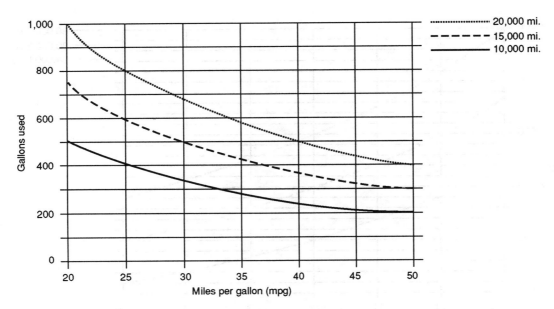

Figure 19–4. Annual automotive gasoline consumption: miles per gallon vs. gallons used.

2. Suppose you drive a 35-mpg car 15,000 miles a year and pay $1.00/gal., whereas your friend's car gets 20 mpg and uses the same kind of gas. How much more would a year's worth of gas cost your friend?

$_____

3. Now suppose your friend's car needs higher octane gas that costs $1.25/gal. How much more would the friend's gas cost per year?

$_____

4. How much more gas does your friend burn in a year?

_____

5. Assuming that pollution is proportional to the amount of fuel burned, how much more pollution does your friend produce, all other things being equal?

_____%

20

Solid Waste Prevention and Management

EQUIPMENT NEEDED

None

People dispose of personal solid waste, material for which they have no use, by littering and by utilizing their municipal waste management system. They drop cigarette butts, candy wrappers, and film packaging wherever they happen to be because they are either too lazy or too uncaring to dispose of them in a more appropriate way. Each of us can easily avoid contributing to that particular type of solid waste problem.

We can also affect the disposal of solid waste generated in the extraction of resources, in industry, and in agriculture (see Fig. 20-1) by supporting regulating legislation. Because the waste is generated by those in business, some of it will be voluntarily recycled because it is obvious that it makes good business sense to do so. We can have the greatest effect on solid waste generated in these sectors by reducing the amount of Earth's resources we use and by getting more use out of those resources we use through reuse and recycling.

Although municipal waste—the waste we contribute to directly through our homes, stores, and offices—amounts to only about 1.5% of the solid waste produced in the United States in a year, it still is a significant amount. It is estimated that the United States produced 185 million tons (370,000,000,000 pounds) of municipal waste in 1990. We can directly help to make these numbers smaller by reducing our use and by reusing and recycling. We also need to look at ways to use waste that is not reusable or recyclable.

Much of the "waste" that people in cities and towns throw away is potentially of some value. Figure 20-2 shows the makeup of municipal waste. The value of some of these materials may not be obvious, but almost all of it can be used in

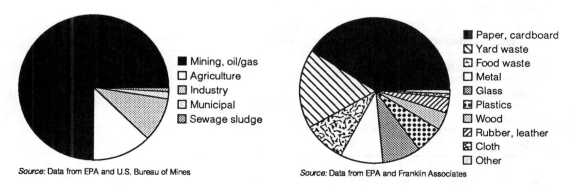

Source: Data from EPA and U.S. Bureau of Mines

Source: Data from EPA and Franklin Associates

Figure 20–1. Solid waste in the United States.   **Figure 20–2.** U.S. urban waste, 1989.

some way. Much of our waste material is packaging material: paper and cardboard, metal, glass, and plastics. Is all of this packaging necessary?

Think of what you throw away. What were the items in your trash originally used for? Look in your trash can and evaluate your trash. Then think about how you can generate less waste, and whether your lifestyle would be appreciably lessened if you were to do so. Also, you can think about whether any of your trash could be reused or recycled. The following table provides space to note the kind of waste you generate. Fill it in, being sure to include things that you might ordinarily overlook. For example, in order to purchase four wood screws, we have to buy ten or twelve wood screws, a plastic bubble, and a piece of cardboard.

EXERCISES

1. Complete the following chart to report on the solid waste you and your family generate.

Waste Category	Items Included	Possible Uses
Paper, cardboard		
Yard waste		
Food waste		
Metal		
Glass		
Plastics		
Wood		
Rubber, leather		
Cloth		
Other		

2. Using the estimate above of 185,000,000 tons, complete the following table to indicate how much of each category of "waste" material is included and how it might be used.

Waste Category	Percent	No. of Tons	Possible Uses
Paper, cardboard	41%		
Yard waste	18%		
Food waste	9%		
Metal	9%		
Glass	8%		
Plastics	7%		

Seeing the actual amounts, rather than percentages, gives us a better idea how much waste there is.

There are several ways of disposing of solid waste. Before 1976, a large part of the municipal solid waste generated in the United States was put in dumps, which were simply hillsides, ravines, or other low or sloping places where people would throw their rubbish. Sanitary landfills replaced dumps, for the most part, sometimes on the same site as the former dump. These have eliminated most of the problems that dumps posed, but they continue to endanger the ground water. Furthermore, they make potentially useful materials unavailable by storing them in the ground. Incineration has become more popular, and recycling is becoming increasingly popular. Each method of solid waste disposal—landfilling, incineration, and recycling—may have certain advantages and disadvantages that depend on the local situation.

A group of Detroit suburbs, consisting mostly of white-collar and professional residents, had a population of 50,200 in 1991. The operator of the incinerator to which their waste was sent estimated that the total solid waste generated in 1991 weighed 32,798 tons.

3. What was the number of tons per person?

4. What was the number of pounds per person?

5. What was the number of pounds per person per day?

6. How does your answer to question 5 compare with the U.S. national average of 3.6 pounds/person/day?

7. How do the "affluent suburbs" compare with the national average in the following table?

Waste Type	National Average	Affluent Suburbs
Paper	41%	22%
Yard waste	18%	31%
Food waste	9%	12%
Metal	9%	5%
Glass	8%	3%
Plastics	7%	3%
Wood	3%	8%

8. By 1991, drop-off recycling of newspapers had been quite well established. Newsprint, it is estimated, makes up about 50% of the waste paper in this community. Would recycling explain why the amount of waste paper was different than the national average?

9. What might be a reason that yard waste is above average in this type of community?

10. Calculate the percent of each type of waste recycled in the "affluent suburbs" according to the data, using the following table. (pounds recycled/pounds produced x 100)

Type of Waste	Pounds Produced	Pounds Recycled	Percent Recycled
Newsprint	3,444	3,229	_____
Steel cans	755	345	_____
Glass containers	916	850	_____
Plastic	269	200	_____
	(According to incinerator)	(According to recycling co.)	

In 1991, the Greater Detroit Resource Recovery Authority, (the "Detroit Incinerator") processed 526,982 tons of the approximately 700,000 tons collected in the city of Detroit.

11. What percentage of the solid waste produced by Detroit is processed by the GDRRA? (tons processed/tons collected x 100)

In 1991, the GDRRA processed some solid waste from other communities, in addition to the Detroit waste, for a total of 733,544 tons. Let's look at what happens to the waste and prepare a breakdown of waste processing.

When the waste arrives at the GDRRA, it passes through several separation processes listed below, not necessarily in the order that they occur in the processing sequence, with the amount of material separated in 1991.

Process	Tons
Separation of bulky waste, such as appliances, mattresses, carpeting	17,228
Recovery of ferrous (iron-containing) metals	38,745
Separation of process residue that passes through a 1-inch screen (Includes "dirt," grass, glass bits, etc.)	142,234

After these materials are separated, the remaining material is incinerated.

12. How much of the total is incinerated?

_____ tons

13. Calculate the percentage of the total for each category:

bulky waste   _____%

ferrous metals   _____%

process residue   _____%

incinerated   _____%

Incineration of the combustible material leaves 127,260 tons of ash.

14. What percentage of the total waste remains as ash?

_____%

15. What happens to the rest of the matter in the waste?

Process residue is landfilled. Ash is placed in a monofill landfill (a landfill designed for just one kind of waste).

According to the Detroit Department of Public Works:

The cost of landfilling is currently $40–50 per ton delivered to the fill site. Cost of delivery varies according to items and distance, and consideration is given to whether it is light material such as cotton waste or heavy materials such as stone. The cost of disposal after delivery to the GDRRA facility approximates $70 per ton. This can vary with the amount and type of material delivered.

The GDRRA facility has reduced Detroit's dependence on landfills by 80% and extended the existing landfill life by 5–9 years.

It uses the heat from incineration to produce electricity and steam for heating large downtown buildings. In 1991 this amounted to:

Steam: Equivalent households   4,500 (33,550,000 kwh)

Electricity: Equivalent households   8,000

Appendix I: Units of Measure

Metric-English Conversions

Length

English		Metric
inch	=	2.54 centimeters
foot	=	0.30 meter
yard	=	0.91 meter
mile (5,280 feet)	=	1.61 kilometer

To convert	multiply by	to obtain
inches	2.54	centimeters
feet	30.00	centimeters
centimeters	0.39	inches
millimeters	0.039	inches

Weight

English		Metric
grain	=	64.80 milligrams
ounce	=	28.35 grams
pound	=	453.60 grams
ton (short) (2,000 pounds)	=	0.91 metric ton

To convert	multiply by	to obtain
ounces	28.3	grams
pounds	453.6	grams
pounds	0.45	kilograms
grams	0.035	ounces
kilograms	2.2	pounds

Volume

English		Metric
cubic inch	=	16.39 cubic centimeters
cubic foot	=	0.03 cubic meter
cubic yard	=	0.765 cubic meters
ounce	=	0.03 liter
pint	=	0.47 liter
quart	=	0.95 liter
gallon	=	3.79 liters

To convert	multiply by	to obtain
fluid ounces	30.00	milliliters
quart	0.95	liters
milliliters	0.03	fluid ounces
liters	1.06	quarts

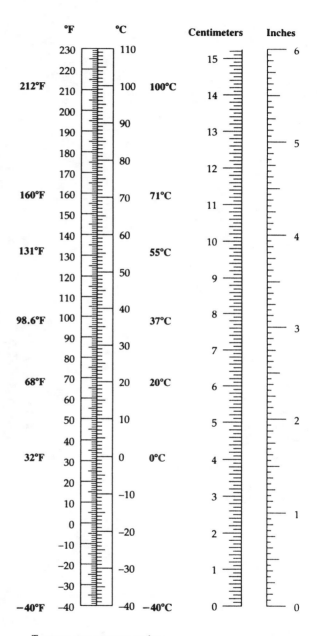

To convert temperature scales:

Fahrenheit to Celsius: $°C = 5/9\,(°F - 32)$

Celsius to Fahrenheit: $°F = 9/5\,(°C) + 32$

Appendix II: Supplies Needed

1 **Introduction to the Compound Microscope**

EQUIPMENT AND SUPPLIES NEEDED

Compound microscopes (1 per student)

Letter "e" slides (1 per student)

Colored thread slides (1 per student)

Assorted slides of instructor's choice (1 per student)

WHERE TO OBTAIN EQUIPMENT AND SUPPLIES

Microscopes (Carolina Biological Supply)

Letter "e" slides (Carolina Biological Supply)

Colored thread slides (Carolina Biological Supply)

2 **Biological Classification**

EQUIPMENT AND SUPPLIES NEEDED

None required

3 **The Plankton Community**

EQUIPMENT AND SUPPLIES NEEDED

Compound microscopes

Prepared slides: *Paramecium caudatum* and *Spirogyra* sp. (1 per student)

Microscope slides and coverslips (2 per student)

Living cultures of *Paramecium* and *Spirogyra* sp. (amounts recommended
by bio. supply co.)

3   The Plankton Community (*continued*)

Methyl cellulose (1.5% liquid) or other mobility inhibitors (± 15 ml bottle per student)

Yeast powder stained with Congo Red dye (5 g per class of 30)

Toothpicks (± 5 per student)

Dilute acetic acid (1%) (± 10 ml bottle per student)

WHERE TO OBTAIN EQUIPMENT AND SUPPLIES

Prepared slides: *Paramecium caudatum* and *Spirogyra* sp. (Carolina Biological Supply)

Microscope slides and coverslips (Carolina Biological Supply)

Living cultures of *Paramecium* and *Spirogyra* sp. (Carolina Biological Supply)

Protozoan mobility inhibitors, Protoslo or methyl cellulose (Carolina Biological Supply)

Acetic acid, 10% (Carolina Biological Supply)

4   Ecological Succession

EQUIPMENT AND SUPPLIES NEEDED

Field guides to plants and animals

35mm slide series on Primary and Secondary succession

WHERE TO OBTAIN EQUIPMENT AND SUPPLIES

Field guides to plants and animals (Carolina Biological Supply)

Succession, 35mm transparencies (Carolina Biological Supply)

5   Exponential Growth

EQUIPMENT AND SUPPLIES NEEDED

None required

6   Population Control

EQUIPMENT AND SUPPLIES NEEDED

Compound microscope (1 per student preferred)

Prepared slides of ovaries from hamster or other small mammal
(1 per student)

WHERE TO OBTAIN EQUIPMENT AND SUPPLIES

Prepared slides: immature, Graafian follicles, corpus luteum
(Carolina Biological Supply)

7   The Bacteria: Representatives of Kingdom Monera

EQUIPMENT AND SUPPLIES NEEDED

Nutrient agar petri plates (1 plate per 4 students, quadrant method)

Sterile cotton swabs (1 per student)

Soil and some type of living insects

Glass slides (1 per student)

Bunsen burner (2 per class)

Methylene blue stain in staining dish

Incubator (32°-35° C)

Compound microscopes (1 per student or as available)

WHERE TO OBTAIN EQUIPMENT AND SUPPLIES

Nutrient agar (Carolina Biological Supply)

Disposable petri plates (Carolina Biological Supply)

Methylene blue stain (Carolina Biological Supply)

Staining dish (Carolina Biological Supply)

Incubator (32°-35° C) (Carolina Biological Supply)

8   Water Quality Testing I: The Coliform Test

EQUIPMENT AND SUPPLIES NEEDED

Select water samples (1 sample per 5 students)

Lactose presumptive test kits (lauryl tryptose may also be used;
    1 kit per student)

BGLB test kits ('demo')

EMB plates ('demo')

Incubator (32°-35° C)

WHERE TO OBTAIN EQUIPMENT AND SUPPLIES

Coliform presumptive test kits (Hach Company)

Brilliant green bile broth (Hach Company)

EMB agar (Hach Company)

9   Water Quality Testing II: Dissolved Oxygen and Biochemical Oxygen Demand

EQUIPMENT AND SUPPLIES NEEDED

Dissolved oxygen test kit (authors use Hach® D.O. test kit;
    1 kit per 2 students)

B.O.D. respirometer (authors use Hach® B.O.D. respirometers; 2 per class)

Applicable water samples (1 D.O. sample per 2 students; B.O.D. 'demo')

WHERE TO OBTAIN EQUIPMENT AND SUPPLIES

B.O.D. respirometer (Hach Company)

B.O.D. chemicals (Hach Company)

D.O. test kits (Hach Company)

D.O. chemicals (Hach Company)

10   Water Quality Testing III: pH

EQUIPMENT AND SUPPLIES NEEDED

Wide range pH (3-10) test kit, colorimetric or pH (authors use LaMotte wide range pH Comparator; 1 kit per 2-3 students)

WHERE TO OBTAIN EQUIPMENT AND SUPPLIES

Wide range pH (3-10) comparator test kit (LaMotte Company)

pH refill for kits (LaMotte Company)

11   Wastewater Treatment

EQUIPMENT AND SUPPLIES NEEDED

In the event of inclement weather, a 35mm slide series on a community sewage plant

WHERE TO OBTAIN EQUIPMENT AND SUPPLIES

If possible, use slides on your own community sewage plant

12   The Human Respiratory System

EQUIPMENT AND SUPPLIES NEEDED

Prepared microscope slides (1 per student)

Vitalometer (respirometer) (1 per ± 15 students)

35mm slide series on lung histology and diseases

WHERE TO OBTAIN EQUIPMENT AND SUPPLIES

Prepared microscope slides of lung tissue (histology) (Carolina Biological Supply)

Wet spirometer for lung exercise (Carolina Biological Supply)

Mouthpiece replacements (Carolina Biological Supply)

Respiratory system, 35mm slide series (Carolina Biological Supply)

13   Air Pollution

EQUIPMENT AND SUPPLIES NEEDED

Pocket combs (1 per student)

Bits of paper, small amount of paper or cigarette ash (per student)

200 ml limewater (per student)

Drinking straws (1 per student)

300 ml beaker (babyfood jar can be used instead) (1 per student)

WHERE TO OBTAIN EQUIPMENT AND SUPPLIES

Pocket combs (students can supply their own or obtain at local store)

Calcium oxide (local garden supply or builders' supply). Prepare the necessary amount of limewater by dissolving calcium oxide (slaked lime) in water until no more will dissolve. Be sure some remains on the bottom of the container to insure the solution is saturated.

Drinking straws (local grocery or party supply store)

Beakers (Carolina Biological Supply, etc.; babyfood jars can be used

14   Thermal Inversion

EQUIPMENT AND SUPPLIES NEEDED

Inversion chamber (1 per class)

Ice

Smoke generator or cigarettes

WHERE TO OBTAIN EQUIPMENT AND SUPPLIES

Temperature inversion chamber (air pollution kit) (Ward's Natural Science)

15   **Introduction to the Arthropods**

EQUIPMENT AND SUPPLIES NEEDED

Preserved specimens of *Procambarus*, any centipede or millipede species, *Romalea microptera*, *Argiope*, *Dermacentor variabilis* or other tick species (1 set per 2 students)

Different insects showing incomplete and complete metamorphosis. Plastic biomounts are useful ('demo')

WHERE TO OBTAIN EQUIPMENT AND SUPPLIES

Preserved insect and arachnid specimens (Carolina Biological Supply)

Insect metamorphosis entomount (Carolina Biological Suply)

16   **Soil Diversity**

EQUIPMENT AND SUPPLIES NEEDED

Soil samples from applicable study sites (1 sample per 2 students)

Berlese-Tullgren funnels (sixteen or 'repeat' extractions preserved in alcohol; 1 funnel per 2 students)

Buchner funnel, optional ('demo')

WHERE TO OBTAIN EQUIPMENT AND SUPPLIES

Berlese funnels (Carolina Biological Supply)

Buchner funnels (Carolina Biological Supply)

Desk lamps (Carolina Biological Supply)

17   Noise Pollution

EQUIPMENT AND SUPPLIES NEEDED

Decibel meters (1 unit per 3-5 students)

Tuning forks: 128; 1,024; 4,096 cps (approximate) (1 set per 3-5 students)

Audio-frequency generator, optional ('demo' for different sound
   frequencies; with generator, author uses small 5" speaker with
   20-20,000 cps frequency range)

Ear model

WHERE TO OBTAIN EQUIPMENT AND SUPPLIES

Sound level meters (Carolina Biological Supply)

Tuning fork set (Carolina Biological Supply)

Ear model (Carolina Biological Supply)

Audio-frequency generator (Newark Electronics or Heath Company)

18   Energy Alternatives

EQUIPMENT AND SUPPLIES NEEDED

Solar energy collector (1 per 2-student group):
   1 shallow wooden box with a piece of window glass or clear plastic
   sheet to cover box *or* 1 one-gallon widemouth glass jar with cover,
   such as used by restaurants. *If using a box*, seal gaps between bottom
   boards with a piece of plywood, masonite, or cardboard, or use duct
   tape; paint the interior surfaces with flat black paint or line with
   black cloth or construction paper; cover top with glass, plastic, or
   food wrap. *If using a jar*, paint or cover half of its interior from bottom to top; make a hole in the cover for thermometer, if necessary.

Black cloth, black paper or black paint (per 2-student group)

Thermometer (per 2-student group)

Solar energy concentrator (1 per 2-student group):

1 solar energy concentrator <u>or</u> 1 set of parts for concentrator: 2 pieces
   of wood cut in parabolic shape and drilled according to pattern;
   1 piece of aluminum sheet; 6-8 tacks or small screws; 14 inches of
   coat hanger wire with paint sanded off. Solar energy concentrators
   should be constructed before the lab. Construction might be done
   in a shop class or as a parent-student project.

WHERE TO OBTAIN EQUIPMENT AND SUPPLIES

Solar energy collector:
Wooden box (grape crates from local fruit market or soda pop cases from local beverage distributor, or boxes could be constructed easily, if necessary)

Box cover (glass or plastic from local hardware store or builders' supply, or food wrap from local grocery)

Widemouth jars (school cafeteria or local restaurant)

Paint, cloth, or construction paper (local hardware store or builders' supply, local variety store, or school art department)

Thermometer (Carolina Biological Supply, etc.)

Solar energy concentrator (materials for construction are readily available at local lumber yards and hardware stores)

19 Energy Conservation

EQUIPMENT AND SUPPLIES NEEDED

None required

20 Solid Waste Prevention and Management

EQUIPMENT AND SUPPLIES NEEDED

None required

Carolina Biological Supply Company
Burlington, NC 27215
Phone (800)334-5551
Fax (503)656-3399
or
Gladstone, OR 97027
Phone (800)547-1733
Fax (503)656-4208

Hatch Chemical Company
P.O. Box 389
Loveland, CO 80539
Phone (800)227-4224
Fax (303)669-2932

Heath Company
Benton Harbor, MI 49022
Phone (800)253-0570
Fax (616)982-5577

LaMotte Company
P.O. Box 329
Route 213 North
Chestertown, MD 21620
Phone (800)344-3100

Newark Electronics
110 South Alpine Road
Rockford IL 61108
Phone (815)229-0225
Fax (815)229-2587

Ward's Natural Science Establishment, Inc.
P.O. Box 92912
Rochester NY 14692
Phone(800)962-2660
Fax (800)635-8439